Automated Guided Vehicle Systems

Günter Ullrich • Thomas Albrecht

Automated Guided Vehicle Systems

A Guide - With Practical Applications -
About The Technology - For Planning

Second Edition

Günter Ullrich
Voerde, Germany

Thomas Albrecht
Fraunhofer-Institut für Materialfluss und Logistik (IML)
Dortmund, Germany

ISBN 978-3-658-35389-6   ISBN 978-3-658-35387-2   (eBook)
https://doi.org/10.1007/978-3-658-35387-2

Translation of the 3. German original edition published by Springer Vieweg, Wiesbaden, Germany, 2019

This Springer Vieweg imprint is published by the registered company Springer Fachmedien Wiesbaden GmbH,
part of Springer Nature.
The registered company address is: Abraham-Lincoln-Str. 46, 65189 Wiesbaden, Germany

Preface

In the 1950s, the automated guided vehicle system (AGV System, AGVS) was invented, which has developed into a proven organisational tool in modern intralogistics to this day. There is practically no industry that does not use automated guided vehicles (AGVs) or at least could do so. From small systems with few or even only one vehicle to systems with well over 100 vehicles, everything is possible and has already been realised. After many years of restraint, companies in the automotive industry are once again the dominant user sector, but many other companies are also using AGVs to optimise their material flows within the framework of Industry 4.0 concepts. This guide illustrates how diverse the applications are and which technological standards are available without claiming to be complete. In addition, we document the new developments that enable innovative application scenarios and open up additional attractive markets. The future has long since begun with the 4th AGVS epoch...

Another focus is the holistic planning of such systems, which is described in detail with all planning steps. Here, the reader will not only find a roadmap through the planning process but certainly numerous valuable hints and clues.

The book authors: Thomas Albrecht and Günter Ullrich

The VDI Technical Committee "Automated Guided Vehicle Systems" has been supporting the industry for over 30 years. Today, it unites about 40 member companies—the European AGV community Forum AGV was created from this strong network, which carries out committed public relations work and, for some years now, has also been providing AGV planning and consultancy with a competent team. We would like to take this opportunity to thank all the members of the Forum AGV, whose contributions have made this primer possible. We would also like to thank Springer Vieweg-Verlag's mechanical engineering department for their kind and understanding support.

The guide is aimed at experts and practitioners in intralogistics who deal with the optimisation of material flows. They are active in almost all branches of industry, in some service companies, or in research and teaching at universities and technical colleges. From our work as planners and consultants, we know that there is a need for a summarising presentation of our topic in practice and in teaching. We have endeavoured to provide an objective view, moderate professional depth, and clear and comprehensible language.

This third edition has been completely revised, is structured slightly differently, and takes into account the rapid developments in technology and markets. For the first time, Dipl.-Ing. Thomas Albrecht is a co-author. He has been working for almost 30 years at the Fraunhofer IML in Dortmund as an AGV specialist and is known in the industry as a reliable and fair-minded authority. May this revised guide contribute to ensuring that automated guided vehicles are used according to their capabilities and become even more efficient in the future.

Voerde, Germany   Günter Ullrich
Dortmund, Germany   Thomas Albrecht
December 2021

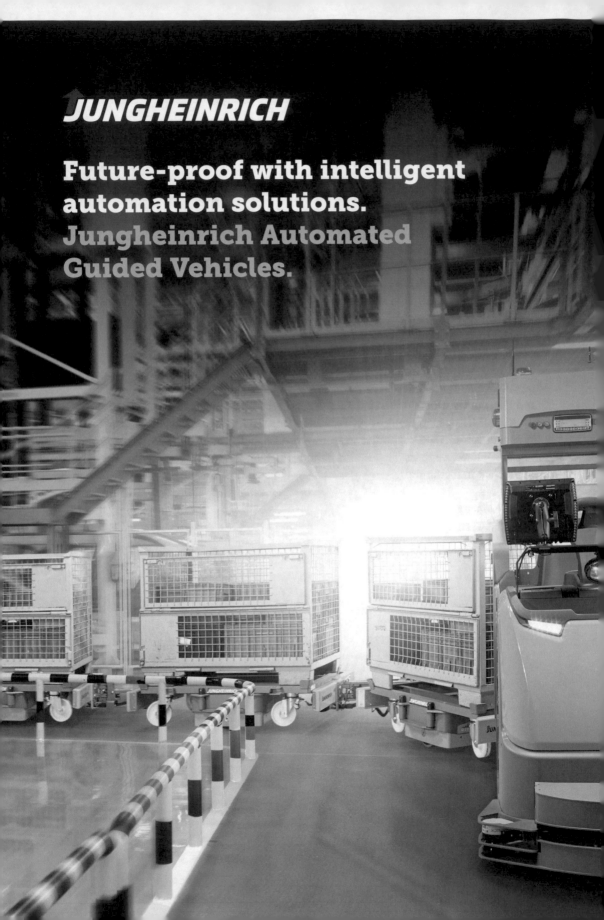

JUNGHEINRICH

Future-proof with intelligent automation solutions.
Jungheinrich Automated Guided Vehicles.

Contents

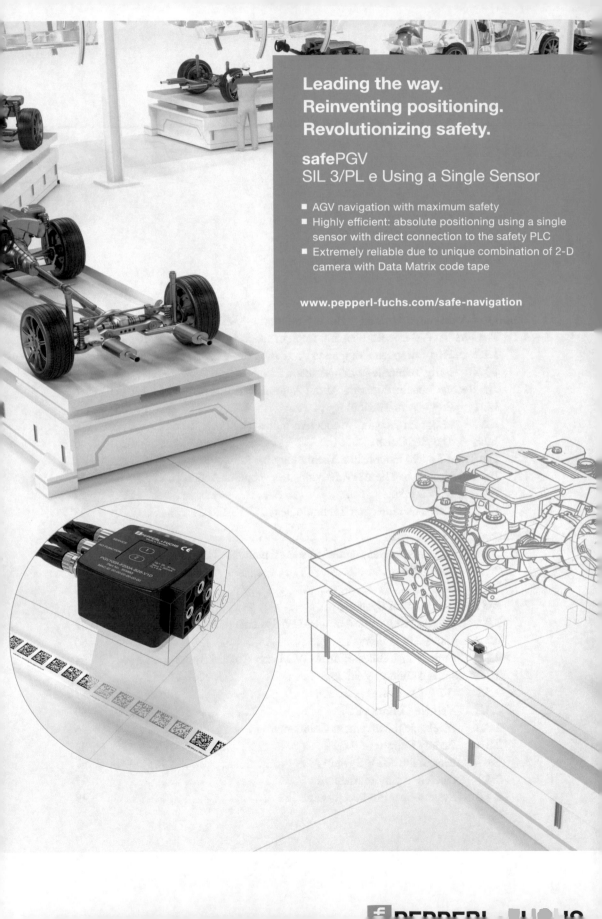

Partner for the **production of the future**

BÄR

AGV / AMR for car transport in final assembly

A changeable final assembly line with autonomous automated guided vehicles sets technological standards in terms of omnidirectional drive technology, patented energy concept with boost caps and an low profile design with integrated lifting table.

AGV / AMR for logistics

Logistics processes such as material supply to automatic stations can be implemented very easily, flexibly and adaptably using FTF / AMR solutions from BÄR Automation GmbH. They can be expanded as required and can be easily integrated into existing or dynamic production structures. All our AGV / AMR can be equipped with area-moving drive units and thus make a major contribution to flexible production of the future.

AGV / AMR for different production requirements

We can offer individual AGV / AMR solutions with a range of standardised vehicles for production, logistics and assembly processes, we can also offer bespoke solutions for customers when required. Assembly processes can be carried out on the AGV / AMR either in stop and go mode or a continuous flow mode. We can utilise all major system interfaces such as VDA5050.X or Siemens SIMOVE. In addition to track-based basic navigation, our AGVs / AMRs can also be equipped with free navigation. We comply with all safety-relevant regulations and protocols and can establish safety-related communication with other systems.

BÄR Automation GmbH · 75050 Gemmingen / Germany
www.baer-automation.de

About the Authors

Günter Ullrich was born in Oberhausen in 1959 and studied general mechanical engineering at the University of Duisburg. There he worked first as a student and then as a scientific assistant in the department of production engineering[1] of Prof. Dr.-Ing. Dietrich Elbracht, who brought with him the subject of AGVs and robotics from his former employer, Jungheinrich AG. During his time at university, Dr. Ullrich dealt scientifically with AGVs and mobile robots. In 1986, Prof. Elbracht founded the VDI Technical Committee for AGVS; Dr. Ullrich was a founding member and has headed the group since 1996.

After his time at the university, Dr. Ullrich was managing director of two companies that planned and sold AGVs and conveyor systems worldwide.

Since 2002, Dr. Ullrich has been an independent AGV planner and consultant in intralogistics. He heads the VDI Technical Committee for AGVS and founded the Forum AGV in 2006. Today, the Forum AGV is known as a constant in the AGV world and, as a community of interests in the AGV sector, is committed to an honest image of the AGV and successful AGV projects. With five competent colleagues, the Forum AGV works very successfully in planning and consulting, primarily for AGV users, but also for companies that (want to) act as suppliers of systems, components, or services in the AGV sector.

Dr. Ullrich wrote about 150 technical papers on the subject of AGVS/mobile robotics.

Thomas Albrecht was born in 1964 in Soest and studied electrical engineering at the TU Dortmund University, specialising in communications engineering. Already during his studies, he worked as a student assistant at the Fraunhofer Institute for Material Flow and Logistics IML (which was then still called Fraunhofer Institute for Transportation Technology and Goods Distribution ITW) on tasks in automation technology and on robot controls. After completing his studies, he became a research assistant at the Fraunhofer IML in 1990 and has been working on all aspects of automated guided vehicle systems

---

[1] Jungheinrich AG was one of the first AGV manufacturers in Europe, and they were also suppliers of industrial robots.

since then: first in software development for vehicle control and tools for driving course programming, then in the development of navigation systems for AGVs, and later as project manager in numerous AGV development projects, as planner and consultant in AGV projects in Germany and abroad, as speaker at conferences and trade fairs, as long-standing active member of the VDI Technical Committee on AGVS, and last but not least as organiser of the AGVS conference, which has been held at Fraunhofer IML in Dortmund since 2012.

Thomas Albrecht is the author of numerous technical publications and co-owner of several patents on navigation procedures and other innovative solutions in the field of AGVs.

NODE.OS
Empowering heterogeneous AMR fleets

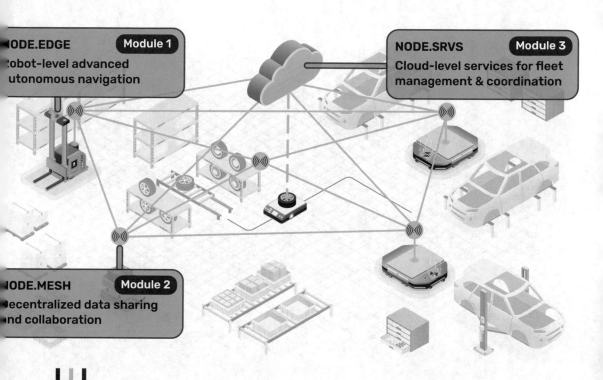

NODE.EDGE — Module 1
Robot-level advanced autonomous navigation

NODE.SRVS — Module 3
Cloud-level services for fleet management & coordination

NODE.MESH — Module 2
Decentralized data sharing and collaboration

A holistic software solution -
operational for all your applications and AMRs.

With NODE.OS, NODE Robotics offers a holistic software solution - from order management to AMR control - versatile in use according to the requirements of your application.

- Industrial-proof: Developed for and tested in industrial environments.
- One software fits all: Modularly applicable without hardware modifications.
- Future-proof: Based on the latest approaches in the field of cloud robotics and machine learning.

Contact us!

Visit us!

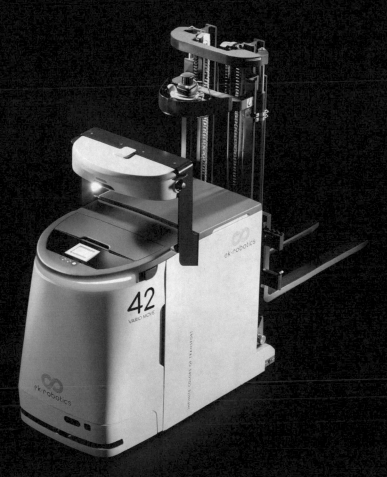

History of Automated Guided Vehicle Systems

1

Summary

Automated Guided Vehicle Systems (AGVS) are an important component of intralogistics. The technological standard and the experience with this automation technology that is now available have led to AGVS finding their way into almost all industries and production areas. The history of AGVS began in the USA in the mid 1950s.

When production started up again after the Second World War and the global economy boomed, automatically moving transport vehicles were part of mankind's dream of making its own work done by machines. The rapid development of sensor and control technology and originally of microelectronics paved the way for AGV systems.

At this point we would like to pay tribute to the invention of the AGV in America, but then concentrate exclusively on the European market. So far there have been few successful American attempts to enter the European market. The opposite approach has been more successful: there are a number of European AGV manufacturers who are carrying out projects in America. The Asian market has had virtually no overlap with Europe in the past, neither in one direction nor the other.

For about 5 years now, China has been experiencing a huge AGV boom, both on the user and, in particular, on the supplier side: in just 2 years (since 2016) the number of Chinese AGV manufacturers has risen from under 10 to over 40. These companies rely both on technology developed in-house and on solutions licensed from European or American suppliers. At present, however, vehicles from Chinese production have not yet appeared on the European market.

The 60 AGV years to date can be divided into four epochs. These are characterised by the technology available and the emotional attitude towards the systems. These epochs can

© Springer Fachmedien Wiesbaden GmbH, part of Springer Nature 2023
G. Ullrich, T. Albrecht, *Automated Guided Vehicle Systems*,
https://doi.org/10.1007/978-3-658-35387-2_1

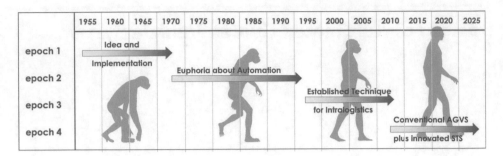

Fig. 1.1 Automatic Guided Vehicle Systems develop in and on evolutionary stages (epochs)

also be understood as stages of evolution, during which there were only limited technical developments and which then merged rather abruptly (Fig. 1.1).

1.1   The First AGVS Epoch: Idea and Implementation

The first epoch began in America in 1953 with the invention of automated transport vehicles and in Europe a few years later. It lasted almost 20 years. Technologically, the first systems were characterised by the simplest track guidance techniques and tactile sensors, such as bumpers or emergency stop bars for workers' protection and safety, with mechanical switches.

At the beginning of the 1950s, an American inventor had the idea of replacing the human being on a towing wagon used to transport goods with an automatic machine.

This idea was implemented by the Barrett-Cravens of Northbrook, Illinois (now Savant Automation Inc., Michigan). In 1954, the Mercury Motor Freight Company in Columbia, South Carolina, installed the first AGV system as a towing train application for recurring consolidated transport over long distances (Fig. 1.2).

The previously rail-guided vehicles now followed an alternating current conductor which was laid in the ground. We know this principle today as inductive track guidance or wire guidance, a vehicle using such technology is called wire guided AGV. The first vehicle thus oriented itself during its journey without a driver by means of an antenna that detected and evaluated the field surrounding the current-carrying conductor with the help of two small coils. The stations at which loads (goods) were to be transferred were coded by magnets embedded in the ground, which were detected by sensors in the vehicle. The coding itself resulted from a specific arrangement of north/south pole oriented magnets.

At that time, the simple control system was based on tube electronics, which had only limited development potential.

Fig. 1.2 One of the first American AGV, built from 1954 as a tractor for five trailers. (Source: Barrett-Cravens/Savant Automation, 1958)

1.1.1   First European Companies

In England the company EMI entered the market in 1956. The vehicles followed a colour stripe on the floor, which was detected by an optical sensor and provided the corresponding control and steering signals. From the 1960s the first transistor-based electronics were used, which increased the options of guidance and control.

In Germany, the companies Jungheinrich, Hamburg, and Wagner, Reutlingen, started the development of AGVs in the early 1960s. They automated the forklift and platform trucks originally designed for manual operation.

The mechanical engineering company Jungheinrich was founded in 1953 and started selling the electric four-wheel forklift truck "Ameise 55" on the market. Then, just a few years later, in 1962, the first automatically controlled, inductively guided stacker "Teletrak" was presented. Optical track guidance was also used here (Fig. 1.3).

The company Wagner Fördertechnik began marketing AGVs for use in automobile production and trade in 1963.

1.1.2   Early Technology and Tasks

Even the first systems, developed and built in the USA, England, Germany and other countries, had elementary features which are still part of an AGV system today: the master

Fig. 1.3   "Ant"/teletrack. (Source: E&K 1965)

control system, the vehicle with control computer, the safety system, and the track guidance system.

The environment in which the first Automated Guided Vehicles were operated was a normal factory or warehouse. Where workers used to use their (towing) vehicles to transport goods through the hall areas, the environment has now been adapted step by step to the requirements of a system that dispensed with human escort. Markings, road ways free of obstacles and passive and active protective measures were to reduce the risks. There is said to have been resistance to the new technology in the USA: trade unions feared that jobs would be lost. But who calculated at the time the gain in new jobs in the developing manufacturing and supply market?

From the mid-1960s onwards, the first individual transport applications and transport as part of the "linking" of workstations were found, and the first systems were used in order picking in the food industry. The variety of vehicles was limited to tractors, forklift trucks and platform trucks (Fig. 1.4).

The master control system was simple, more precise: there wasn't any. The vehicles drove predetermined roadways from station to station, started on demand and stopped after recognizing the stop markers. A simple electrical system and a magnetic sensor system made this possible in a reliable manner. The operation did not allow any flexibility; the transport bridged farther distances, the stations were approached one after the other, there was practically only one direction—forward.

The AGV had been developed based on a manually operated towing vehicle, i.e. it had the same steering and drive as a normal vehicle, and it had additional safety devices. Its size

Fig. 1.4 Ant/teletrak with trailers. (Source: E&K)

was determined by the application requirements. If the driver was removed, then a combination of mechanics, electrics and "electronic intelligence" had to take over his tasks. Human perception—through the driver's eyes—was thus replaced by sensor technology, albeit only in a rudimentary form. In order to ensure safety in operational traffic, not only the equipment had to be protected, but above all the people working in the company.

Initially, the vehicles' control system was still based on tube technology, then there were those with relays and step switching mechanisms, and from the late 1960s on, semiconductor technology (TTL logic) came up.

Personal safety for forward direction was realised with a "bumper" or a safety edge, i.e. in any case with a tactile sensor.

The track guidance was provided by current-carrying conductors in the floor or by optical guidance lines on the floor.

At the end of the 1960s the first tractors with automatic couplings were designed: they could pull one or more trailers and park them (= automatically uncouple them) where they were needed. However, coupling and reversing was still done manually by an operator using the fold-down drawbar. The following picture shows such a tractor, interesting here is also how unsecured the attached trailer was (Fig. 1.5).

Fig. 1.5 AGV as towing tractor. (Source: E&K approx. 1965)

1.2   The Second Epoch: Euphoria About Automation

The second epoch lasted through the 1970s and 1980s and ended in the early 1990s. The electronics arrived in the form of simple on-board computers and large control cabinets for the block section control of the plant. Active inductive guidance by means of a wire in the ground became commonplace, and data transmission was either via the same wire, infrared or even already by radio.

In the 1970s, the typical AGV was finally created. As production efficiency increased and manually operated transport systems were used, the demand for a higher degree of automation developed, which should reduce production costs in the long term.

1.2.1   Progress in Technology

Market demand, driven by user expectations, could only be satisfied by constantly improving technology.

A growing number of manufacturers and component developers increased the flexibility of the application possibilities and improved the system capabilities. Even at this early stage, the manufacturers recognised that they could take advantage of the rapid developments in electronics and sensor technology. However, a special supplier market did not develop; the overall market volume was too small for this. Developers and manufacturers of components were driven by other markets, e.g. by the needs of manufacturers of traditionally manned transport vehicles.

The experience of AGV manufacturers was increasingly incorporated into improved plant control systems. But the supplier community still had its roots in mechanical engineering.

Technical innovations freed the manufacturers from previous restrictions, and a number of innovations came onto the market in the 1970s:

- Powerful electronics and microprocessors enabled increased computing power and thus more complex application scenarios and plant layouts. Programmable logic controllers (PLC) were used for the first time in plant control. Improved, affordable sensor technology improved precision while driving, navigation (positioning and position recognition) and at the load transfer stations.
- Battery technology became more powerful, although one had to admit in retrospect that it was not fully mastered. Automatic battery charging was also introduced.
- One navigation method became generally accepted: inductive track guidance, also known as wire guidance. An alternating current flowing through a conductor in the ground generates an alternating magnetic field around the conductor, which in turn induces a voltage in a coil, the level of which depends on the position of the coil relative to the conductor. If two coils are arranged below the vehicle in such a way that one is to the left and one to the right of the conductor wire, the differential voltage of the two coils can be used to control the steering motor.
- The master control system was modelled similar to the block control system of railway traffic. Large electrical cabinets using relay technology provided for sequence control and ensured that the vehicles did not collide or block each other.
- Load handling was done more intelligently and increasingly automated. The possibilities for movement of the vehicles increased (reverse travel with load transfer, omni-directional movements); the first outdoor applications were realised.
- The automated vehicles were fully integrated into production processes; for example, the vehicles were used as mobile workbenches (series assembly).
- For data communication, infrared light but also radio systems were used.

1.2.2 Major Projects in Automotive Industry

Market demand was mainly driven by the automotive industry. The large German car manufacturers in particular modernised and automated seemingly limitlessly. The AGV was one of them, it was "in", especially in the following application areas:

Fig. 1.6 Wire guided assembly platforms for motors at VW in Salzgitter (Source: E&K 1977)

- taxi operation in intralogistic applications,
- AGV as a mobile workstation in the pre-assemblies,
- interlinking of production machines in aggregate manufacturing,
- tractors, piggyback and forklift truck AGVs for assembly belt supply,
- in the warehouse, for picking and material delivery to the lines,
- special devices for integration into production systems.

Many of the major VDI-FML[1] partners in the automotive industry have supplied large systems, often with more than a hundred vehicles. The systems were used in pre-assembly (cockpit, front end, doors, engines, transmissions, drive trains), in final assembly, in buddy shop, but also for logistic tasks (Figs. 1.6, 1.7, and 1.8).

1.2.3 The Big Crash

By the end of the 1980s, the decline was already looming: The economy was hit by a recession and money became scarce. AGVs had the image of being expensive anyway: The flexibility the systems were advertised with even then was not achieved in practice. Small changes in the drive track had to be carried out by the AGV supplier and cost a lot of money. The reliability and availability of the systems left much to be desired.

The German car manufacturers Volkswagen, BMW and Mercedes Benz agreed that something had to be done about the lack of compatibility and economy of the systems. They initiated the foundation of the VDI technical committee[2] "Automated Guided Vehicle Systems", which began in 1987 to develop VDI guidelines on the relevant AGV topics

---

[1] VDI FML (= Fördertechnik, Materialfluss und Logistik)—German for: Conveying technology, material flow, logistics.

[2] www.vdi.de

Fig. 1.7 Car production with AGVs: Body shop of the VW Passat at VW in Emden. (Source: DS AUTO MOTION 1986)

Fig. 1.8 Power train pre-assembly at VW in Hannover. (Source: DS AUTOMOTION 1986)

under the chairmanship of Duisburg University professor Prof. Dr.-Ing. Dietrich Elbracht. Since 1996 the group has been headed by Dr.-Ing. Günter Ullrich, who was also a founding member.

Four years later, this VDI technical committee then held the first AGV conference[3] in Duisburg, where these topics were intensively discussed. In addition, the Forum AGV,[4] the European AGV community in which the main AGV manufacturers in Central Europe (Finland, Belgium, the Netherlands, Germany, Austria, Switzerland) are organised, was created in 2006.

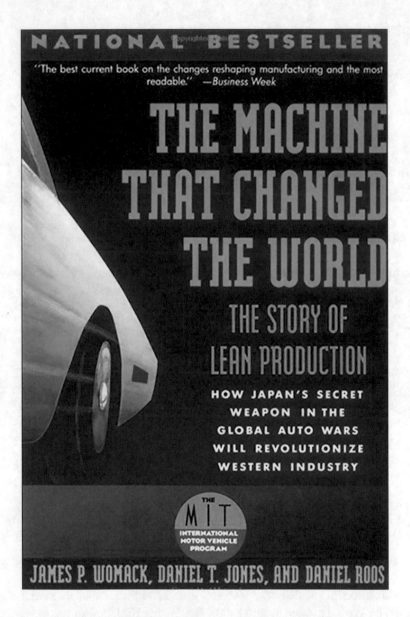

---

[3]The AGV conference is held every two years since 2012 at the Fraunhofer Institute IML in Dortmund (www.fts-fachtagung.org), from 2002 to 2010 at the University of Hanover and at the beginning, i.e. in 1991, 1993, 1995, 1998 and 2000 at the University of Duisburg.

[4]www.forum-fts.com, where you will also find a list of the currently successful European AGV manufacturers.

However, the AGV sector could not escape the temporary decline, mainly caused by an American book (Fig. above), which contains an MIT study[5] on the productivity of the global car manufacturers. This study stated that Japanese car manufacturers were able to deliver better quality at lower manufacturing costs with the simplest of means and new labour structures.

This study led to a complete rethink in Europe. For the large AGV systems it meant the end of the AGV recession. Many "big" AGV manufacturers ended their involvement in AGVS or took the path of licensing in a more global world. But in the end, a new start was on the cards with new medium-sized players, new technology, new products and new customers (new industries)!

1.3   The Third Epoch: Established Technology for Intralogistics

The third epoch lasted from the mid-1990s until around 2010, during which time technological standards were created and markets consolidated. The vehicles have electronic controls and contactless sensors. A standard PC is used as AGV master control system, with either a PLC or a microcomputer onboard the AGV. Line guidance is no longer important, the "free" navigation techniques of magnet and laser navigation are becoming more and more popular. WiFi is establishing itself as a data transmission technology.

This epoch is characterised by the fact that the forefront of the automotive industry has been broken by a multitude of different users. The number of AGV units per system is no longer as large as in the second epoch. And there is another important feature that distinguishes the AGV for the first time: AGV systems are now reliable, proven means of intralogistics. Manufacturers can choose from plenty of proven technologies which they combine to create reliable, high-performance and approved products.

Advances in material flow and storage technology, improved production methods in mechanical engineering and new trends in assembly techniques support AGV development. Additionally, advanced computer and sensor technologies are bringing further significant advances in vehicle and control technology and in application areas:

- vehicles with increased speed when driving, manoeuvring, load handling thanks to improved sensor technology,
- low-cost vehicles, or better: simple solutions,
- alternative energy concepts with inductive energy transmission,
- new navigation methods (magnetic grid, laser triangulation, RFID transponders, contour navigation),

---

[5] James P. Womack, Daniel T. Jones, Daniel Roos: The Machine That Changed the World: The Story of Lean Production. Publisher: Harper Paperbacks; Editin: Reprint (1st November, 1991).

- the triumphal march of the PC—in the vehicle, in system control and in intelligent sensor technology,
- data transmission now mostly via WiFi,
- new functional areas, such as the operation of a block storage system, in the production department, in the "fractal factory" (lean production) or in a hospital.

In principle, any kind of unit loads can be transported with AGV. All companies in which pallets, lattice boxes, containers, bins, boxes, parcels and the like are transported can generally use AGVs. Thus, since the mid-1990s until today, more and more industries have made use of AGVs, but in contrast to the automotive industry in the second epoch, always with caution and usually with success.

The handling of goods has changed from the original unilateral transport to a multi-dimensional movement, as vehicles now have facilities to move goods from virtually anywhere to anywhere in the warehouse or manufacturing area. In addition, they can position the goods for assembly in an ergonomic way that meets the needs of the customer. Complex transport networks are emerging, with a large number of vehicles and intersecting routes and an ever-increasing number of load transfer stations.

In Japan, following the Kaizen principles, existing supply shelves on the production lines were converted into automated logistics units. For this purpose, a modular AGV construction kit was developed, which combined all the necessary elements of a simple magnetic track guidance system up to the control system in a closed unit.

A new market for the AGV is emerging worldwide with hospital logistics, which is becoming more and more interesting because the AWT systems used to date—such as the EMS or, even earlier, the P&F systems—will be replaced more or less/earlier or later by the AGV.[6]

The technology and applications during this epoch will be the subject of the following sections, so we can be brief here. It is important to have an overview of which AGV manufacturers play an important role today and where they have their roots. Therefore, Table 1.1 lists relevant European suppliers of the AGV market and Table 1.2 gives an overview of the most important stations of the Central European market.

Finally, it should not go unmentioned that, in addition to the industrial companies mentioned, research institutions, e.g. university chairs and Fraunhofer Institutes, have also dealt with the subject of AGVs. The main focus has been on selected and demanding problems in the fields of navigation, control engineering or online path planning. Innovative solutions were also developed which were sometimes far ahead of their time. Examples include the first industrial robot on a mobile platform ("MobiRob" by the University of Duisburg), the first service robot ("Care-O-bot" by Fraunhofer IPA) or the decentralised self-organisation of a larger group of autonomous vehicles ("AGV swarm" by Fraunhofer IML) (Figs. 1.9 and 1.10).

---

[6] Abbreviations (based on German language): AWT—Automatic goods transport system; EHB—Electric Monorail System; P&F—Power and Free (chain conveyor).

NODE.OS

mpowering heterogeneous AMR fleets

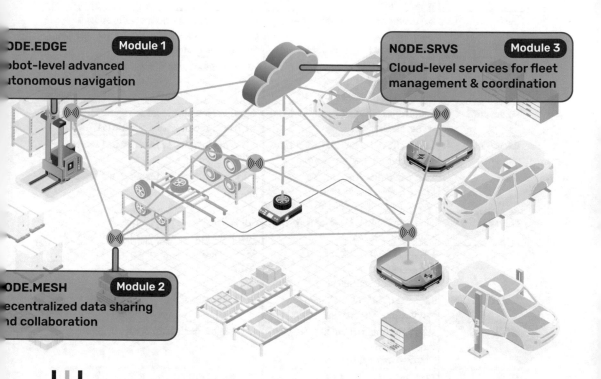

)DE.EDGE — **Module 1**
bot-level advanced
jtonomous navigation

NODE.SRVS — **Module 3**
Cloud-level services for fleet
management & coordination

)DE.MESH — **Module 2**
ecentralized data sharing
nd collaboration

A holistic software solution -
operational for all your applications and AMRs.

With NODE.OS, NODE Robotics offers a holistic software
solution - from order management to AMR control - versatile
in use according to the requirements of your application,

- ✓ Industrial-proof: Developed for and tested in industrial environments.
- ✓ One software fits all: Modularly applicable without hardware modifications.
- ✓ Future-proof: Based on the latest approaches in the field of cloud robotics and machine learning.

Contact us!

Visit us!

Table 1.1 Relevant Central European companies in the AGV sector

Abbreviation	Name	Description
AGILOX	Agilox Systems GmbH, A—Vorchdorf www.agilox.net	AGV manufacturer
ASTI	ASTI Mobile Robots, E—Burgos www.asti.es	AGV manufacturer
BÄR	BÄR Automation GmbH, D—Gemmingen www.baer-automation.de	AGV manufacturer, especially special purpose vehicles for assembly automation, member of Forum AGV
BITO	BITO Storage Technology Bittmann GmbH, D—Meisenheim www.bito.com	AGV manufacturer ("Leo Locative")
BlueBotics	BlueBotics SA, CH—St. Sulpice www.bluebotics.de	Supplier of control and navigation systems for AGV, member of Forum AGV
CREFORM	CREFORM Technology GmbH, D—Baunatal www.creform.de	AGV manufacturer, member of Forum AGV
dpm	Daum+Partner Engineering GmbH, D—Aichstetten www.daumundpartner.de	AGV manufacturer, member of Forum AGV
Dematic	Dematic GmbH, D—Bremen www.dematic.com	Logistics system provider, AGV manufacturer, member of Forum AGV
DS-A	DS AUTOMOTION GmbH, AT—Linz www.ds-automotion.com	AGV manufacturer, member of Forum AGV
ELETTRIC80	Elettric80 S.P.A., I—Viano www.elettric80.com	AGV manufacturer
E&K	E&K Automation GmbH, D—Rose garden www.ek-automation.com	AGV manufacturer, member of Forum AGV
FOX	Department FOX of Götting KG, D—Teacher www.goetting.de/fox	Department of Götting KG, which automates commercial trucks for outdoor use
Götting	Götting KG, D—Lehrte www.goetting.de	Component and AGV accessory manufacturer, member of Forum AGV
Grenzebach	Grenzebach Mechanical Engineering GmbH, D—Asbach-Bäumenheim www.grenzebach.com/de/produkte-maerkte/intralogistik/	Logistics system provider, AGV manufacturer, member of Forum AGV
Guidance Automation	Guidance Automation Ltd, UK—Leicester www.guidanceautomation.com	Supplier of control and navigation systems for AGV
InSystems	InSystems Automation GmbH, D—Berlin www.insystems.de	AGV manufacturer

(continued)

Table 1.1   (continued)

Abbreviation	Name	Description
Jungheinrich	Jungheinrich Moosburg AG & Co KG, D—Moosburg www.jungheinrich.de/systeme/ fahrerlose-transportfahrzeuge	Logistics system provider, manufacturer of industrial fork lift trucks, offers AGVs on the basis of its (own) automated series trucks, member of Forum AGV
Knapp AG	A—Hart near Graz www.knapp.com	Logistics system provider, AGV manufacturer
KUKA	KUKA AG, D—Augsburg www.kuka.com/de-de/produkte-leistungen/mobilitaet	Supplier of mobile robot platforms and AGV navigation solutions
Leuze	Leuze electronic GmbH + Co. KG, D—Owen www.leuze.de	Supplier of safety components and systems, member of Forum AGV
Linde MH	Linde Material Handling GmbH, D—Aschaffenburg www.linde-mh.de	Logistics system provider, manufacturer of industrial fork lift trucks, offers AGVs on the basis of its (own) automated series trucks
Magazine	Magazino GmbH, D—Munich www.magazino.eu	Manufacturer of mobile robots for order picking
MIR	Mobile Industrial Robots A/S, DK—Odense www.mobile-industrial-robots. com	AGV manufacturer
MLR system	MLR System GmbH, D—Ludwigsburg www.mlr.de	AGV manufacturer, member of Forum AGV
Navitec	Navitec Systems Oy, FI—Espoo www.navitecsystems.com	Supplier of AGV track guidance and navigation systems for AGVs
Oceaneering	Oceaneering AGV Systems GmbH, D—Leinfelden-Echterdingen www.oceaneering.com/AGV	AGV manufacturers, member of Forum AGV
Rocla	Rocla OY, FI—Järvenpää www.rocla.com	AGV manufacturers, member of Forum AGV
Schabmüller	Schabmüller GmbH, D—Berching www.schabmueller.de	Components manufacturer for drive and steering motors
SEW	SEW-Eurodrive GmbH & Co KG, D—Bruchsal www.sew-eurodrive.de	Component and system manufacturer for drive and energy technology, AGV manufacturer
SICK	Sick AG, D—Waldkirch www.sick.com	Supplier of safety components and systems, member of Forum AGV
SimPlan	SimPlan Integrations GmbH, D—Witten www.simplan.de	Service provider for simulation and emulation

(continued)

Table 1.1 (continued)

Abbreviation	Name	Description
SSI Schäfer	SSI Schaefer Automation GmbH, D—Giebelstadt www.ssi-schaefer.com	Logistics system provider, AGV manufacturers, member of Forum AGV
STILL	STILL GmbH, D—Hamburg www.still.de	Logistics system provider, manufacturer of Industrial fork lift trucks, offers AGVs on the basis of its (own) automated series trucks
swisslog	Swisslog Holding Ltd, CH—Buchs www.swisslog.com	Logistics system provider, AGV manufacturers, member of Forum AGV
TÜNKERS	Tünkers Mechanical Engineering GmbH, D—Ratingen www.tuenkers.de	AGV manufacturers, especially assembly and special purpose vehicles
WFT	STÄUBLI WFT GmbH, D—Sulzbach-Rosenberg www.wft-gmbh.de	AGV manufacturer, especially special purpose vehicles

The conclusion at the end of the third AGV epoch at the transition to the current fourth epoch is as follows: The implementation of customer requirements has produced a wealth of system designs. To the same extent, the demands of the customers have grown along with them. With the expanded range of applications and technical development, the complexity of the systems has also increased.

Now, with increasing complexity, system costs must not rise as well. The customer expects, similar to the market for consumer products (PC/laptop, smart phone, television etc.), that performance improvements are realised and offered without price increases. The success of optimisation through automated systems today lies in a well coordinated mix of different means of transport and degrees of automation. Manufacturers can draw on their many years of planning experience and proven technologies.

1.4   The Fourth Epoch: The AGV Expands the Scope of Action

The fourth epoch of the AGV is not new in all respects: The contents of the third epoch—technology and basic applications in intralogistics—remain up-to-date and will continue to be so, but new technical possibilities and functionalities and thus new applications, but also challenges, will be added. For the first time, this new epoch will not completely displace the old one, but will build on its achievements. So: the applications and technical solutions that were good in the third epoch are still good today!

If the vintage AGV was, according to definition and lived practice, used exclusively for internal material transport, automatic vehicles today take on additional tasks that can be

Table 1.2 Stations of the current AGV manufacturers, limited to the European market

Year	AGV manufacturer	Event/previous history
1953	All	The Barrett company starts in America with AGVs
1956	All	EMI produces AGVs in England; in Sweden in 1973 VOLVO started to produce AGVs with Kalmar
1962	E&K	Jungheinrich, Hamburg, starts with AGVs (Teletrak)
1963	E&K	Ernst Wagner KG, Reutlingen, starts the development of automatically moving vehicles
1969	Egemin	Egemin delivers the first AGV, based on purchased vehicles, however
1970	Swisslog	Telelift starts with the Transcar AGV. Since 1973 in Puchheim near Munich
1971	MLR	Babcock und Bosch Transport- und Lagersysteme is established in Stuttgart, later (1983) in Schwieberdingen. The nucleus for AGV components (steering control and guidance signal generator) was the "Bosch Transport- und Lagersysteme" division within Robert Bosch GmbH, Stuttgart. Wagner initially (until 1971) installed these parts in his vehicles and systems. Later, the division went to Babcock and then to MLR
1971	E&K	Ernst Wagner KG, Reutlingen, founds the division "Automated Guided Vehicle Systems"
1973	E&K	Mannesmann takes over DEMAG. Mannesmann Demag Fördertechnik AG, Wetter, is founded in 1992. The company's roots go back to 1910, when Deutsche Maschinenfabrik AG (DeMAG) was founded. At the same time, Leo Gottwald KG was founded in 1956, building harbour cranes and later AGVs in the harbour area. In 2006, Demag Cranes & Components GmbH and Gottwald Port Technology GmbH (GPT) merged under the umbrella of Demag Cranes AG, and since Jan. 2017, as Terex MHPS, part of Konecranes, which still manufactures AGVs for use in container ports
1974	MLR	Change of name from Babcock and Bosch Transport- und Lagersysteme to Babcock Transport- und Lagersysteme, as Babcock will take over 100% of the shares
1980	MLR	The Pohling-Heckel-Bleichert AG in Cologne takes over the transport and storage systems division of Babcock. From now on the company is called PHB Transport- und Lagersysteme
1980	E&K	Mr. Eilers and Mr. Kirf found an engineering office for control technology and have been Jungheinrich's system partner since 1988
1983	Rocla	Rocla starts with AGVs in Finland
1984	FROG	FROG starts in NL; initially as Frog Navigation Systems. 2007 then new start as FROG AGV Systems

(continued)

Table 1.2 (continued)

Year	AGV manufacturer	Event/previous history
1984	DS-A	The Austrian conglomerate Voest Alpine AG starts with AGVS. In 1991, after a restructuring, the AGVS branch is assigned to the newly founded VA Technologie AG
1985	MLR	PHB Transport- und Lagersysteme takes over MAFI Transport-Systeme and 1 year later Trepel GmbH, Wiesbaden. In addition, the holding company PHB Gesellschaft für Industriebeteiligungen is formed, including the companies PHB Transport- und Lagersysteme, Eisgruber, Mafi, Trepel and BBT
1986	E&K	Linde acquires a two-step shareholding (1986 and 1988) in Wagner Fördertechnik. With the complete takeover in 1991, the field of AGVS is spun off as an independent company, INDUMAT
1989	Swisslog	Thyssen Aufzüge takes over Telelift
1990	MLR	Noell, Würzburg takes over the AGVS activities of the PHB Group. Name: Noell, Schwieberdingen branch. Noell was part of the Preussag-Salzgitter Group
1993	MLR	Noell takes over the company Autonome Roboter, Hamburg and in 1994 Schoeller Transportautomation, Herzogenrath
1993	E&K	The AGV activities of Mannesmann Demag and Jungheinrich will be transferred to Demag-Jungheinrich FTS GmbH, Hamburg.
1994	DS-A	TMS Transport- und Montagesysteme GmbH is founded under the name of VA Technologie AG, which among other things continues the AGV activities
1996	E&K	Eilers & Kirf takes over Demag-Jungheinrich FTS GmbH
1997	MLR	MLR takes over the AGVS division of Preussag/Noell
1999	Swisslog	Swisslog takes over Telelift from Thyssen Aufzüge. Swisslog emerged from the former Sprecher & Schuh AG (in CH-Aarau since 1898)
1999	CREFORM	Foundation of CREFORM Technik GmbH Germany, subsidiary of Yazaki Industrial Chemical Co. (Shizuoka, Japan) and its US subsidiary CREFORM Corporation (Greer, USA), objective: marketing of simple, flexible, material handling systems (the modular AGV construction kit)
2000	FOX	The Götting KG company founds the independent subsidiary FOX, which automates series commercial vehicles such as trucks and wheel loaders
2000	Egemin	Egemin starts building its own vehicles
2001	E&K	E&K takes over INDUMAT from Linde
2001	DS-A	VA Technologie AG is selling the TMS group to the French conglomerate VINCI. TMS Automotion GmbH will be established there in 2002 to continue the AGV business

(continued)

Table 1.2   (continued)

Year	AGV manufacturer	Event/previous history
2001	BlueBotics	BlueBotics SA, founded in Switzerland, offers AGV manufacturers "ANT" (Autonomous Navigation Technology), a licensable software for contour navigation
2004	Snox	The Snox Engineering Group (France, Belgium) starts the AGV business
2005	DS-A	HK Automotion, Austria takes over the entire TMS Group; in 2008 the name is changed to DS AUTOMOTION
2007	BÄR	BÄR Automation (founded 1972) starts AGV activities, develops and builds customised solutions for automated assembly lines
2008	Götting KG	The activities of Fox GmbH are continued by the parent company Götting KG as a department
2008	MT Robot	Foundation of MT Robot AG in Zwingen, CH
2008	Rocla	Rocla becomes part of Mitsubishi Caterpillar Forklift Europe
2011	Serva	Servapark GmbH (Bernau, D) starts AGV business as a privately financed start-up company; target market: car parking robots (automated car parking done by AGVs)
2012	Swisslog	Swisslog Healthcare Solutions is merging with the JBT Corporation from Chicago (USA) in hospital logistics, getting rid of the name Telelift and focusing its activities in Westerstede. Separation from the small conveyor systems product division and the associated Telelift brand name
2012	Swisslog/ Grenzebach	Grenzebach Group takes an 11.3% stake in Swisslog Holding
2013	Grenzebach/ Snox	Acquisition of Snox by Grenzebach Maschinenbau GmbH
since about 2014	Jungheinrich, STILL, Linde	The major German manufacturers of industrial trucks are re-entering the AGV business after many years of abstinence
2014	FROG	Acquisition of FROG by Oceaneering
2014	KUKA/Swisslog	Takeover of Swisslog Holding by KUKA AG
2014	SSI Schäfer	SSI Schäfer develops its first own AGV for the transport of small boxes/bins and cartons with low weight (named "WEASEL")
2015	BITO	BITO presents Low-Cost-AGV "Leo Locative" for transport of bins, developed in cooperation with Fraunhofer IML
2015	Swisslog/ Grenzebach	Swisslog takes over parts of the AGV activities from the Grenzebach Group.
2015	Magazino	Magazino GmbH is created as a spin-off of the TU Munich; target: mobile picking robots, "robots to goods" 2015 participation of SIEMENS AG; winner of several logistics awards in the following years

(continued)

Table 1.2 (continued)

Year	AGV manufacturer	Event/previous history
2015	ROFA/MLR	ROFA Industrial Automation AG with headquarters in Kolbermoor takes over MLR System
2015	SSI/MoTuM	SSI Schäfer acquires a majority stake in Belgian AGV manufacturer MoTuM
2015	Linde MH/Balyo	Linde and the French AGV manufacturer Balyo agree on strategic cooperation
2016	Dematic/NDC	Dematic acquires NDC Automation (AGV manufacturer in Australia).
2016	Dematic/Egemin	Dematic acquires Egemin and becomes the world's largest AGV manufacturer
2016	KION/Dematic	KION Group takes over Dematic
2018	SSI Schäfer/DS-A	SSI takes a stake in DS AUTOMOTION

Fig. 1.9 The world's first mobile industrial robot MOBIROB. (Source: University of Duisburg, Department of Production Engineering, Prof. Dr.-Ing. D. Elbracht, 1985)

Fig. 1.10 Cellular transportation technology/"AGV swarm" in the Fraunhofer IML test hall in Dortmund (2011)

described as a service or service for people—especially outside of an industrial environment: Providing information in museums, transporting suitcases in hotels or airports, cleaning floors in supermarkets, airport or railway station halls, security inside buildings during the night hours or even offering "helping hands" and assistance in hospitals or in a home for the elderly. There are already numerous application examples for the latter environment in connection with human-like looking mobile robots, especially in Japan. These systems "work" very close to and together with people, to whom they read aloud, have simple conversations with them, give assistance when getting up or even fetch help in case of a fall or sudden health problems.

It is easy to imagine that the challenges that automatic vehicles of this new generation will have to face, and which derive in particular from their close proximity and "cooperation" with people, will give rise to new solutions that will also be of interest for AGVs in industrial environments and will allow a number of new or at least improved applications.

This sets the priorities for the following pages. It is about new and old markets as well as the functional challenges of the fourth epoch. The present is described, a look into the near future, which is still part of the fourth epoch, is then ventured into Chap. 4. For reasons of space and in order not to lose focus, we will not take a closer look at the large and rapidly growing market for consumer products, e.g. hoover and mowing robots for private use.

1.4.1   New Markets

By new markets, we mean applications of automatic driving outside the typical intralogistics in industrial environments, i.e. in semi-public, public or even private areas. A very vivid example of this are AGV systems that are used in hospital logistics (so-called healthcare logistics), as the restricted area *basement/logistics level* with trained personnel and vintage AGV technology, and the semi-public area *hospital ward* with patients and visitors, are only a few centimetres apart, i.e. one elevator door thickness: As soon as an AGV, coming from the basement by freight elevator, leaves the elevator's cabin for delivery or pick up of a transport container and enters the publicly accessible area of a hospital ward, it encounters "people from outside the company" there, which fundamentally changes the technical, organisational and legal situation.

Compared to the typical use of an AGV in an industrial environment where it usually encounters trained, adult and healthy employees, who have a cooperative approach to vehicles, the situation is completely different in a hospital or more generally in the public sector. Here one has to reckon with non-instructed persons, with patients, with visitors of all ages, even with playing children etc. This group of people—they are not users or operators, but encounter the vehicles more or less unexpectedly—can also be found in other areas of public life, such as in a supermarket, DIY store, library, museum or amusement park. From fear to curiosity to destructive rejection, all reactions must be expected here, i.e. a vehicle must be able to cope with all this, or—more precise—its programmers must have developed and installed behaviour and solution strategies for all these situations and behavioural patterns.

Service robots (SR), which have been around for a long time in principle, albeit in very limited numbers, have always had to find their way around in such areas and situations. Until now, however, the AGV and service robotics have had little contact. Today, these product areas have merged and a new type of automatic vehicle has emerged, expressed by the simple formula

$$AGVS + SR = STS$$

If the vintage AGV is crossed with service robotics, service and transport systems (STS) are created. STS are therefore automatic vehicles that not only transport, but can also perform a variety of service tasks or—from the robot's point of view—service robots with transport function. The following pictures show typical examples: Vehicles that can perform a variety of tasks in hospitals, nursing homes or hotels, for example food distribution, fetch and bring services for medicines, drinks or documents, luggage trolleys and signposts; in supermarkets they can help to find specific products (Figs. 1.11 and 1.12).

From the world of AGVs, the new product has benefited from mechanical engineering expertise and can move and carry material. The service robot genes have provided the intelligent MORE:

Fig. 1.11 Current version of the Care-O-bot, application: Providing customers in an electronics store with product information/leading them to the product they are looking for. (Source: mojin robotics)

Fig. 1.12 Three of the first STS products on the market: use in a hospital, retirement home or hotel. (Source: left: MT-Robot, centre: MLR, right: Cleanfix/BlueBotics)

Fig. 1.13 Cleaning robot used in the supermarket. (Source: Nilfisk)

- More technology in object recognition: 3D sensor technology, sensor fusion
- More technology in navigation: improved contour navigation, 3D maps
- More flexibility (simplicity, comprehensibility) during commissioning /changes
- More "autonomy": independent decision-making, e.g. dynamic evasion
- More service-friendliness (information on core components)
- More variants in energy supply: Lithium-ion technology, contactless charging

The cleaning robot is a good example of this. Floor cleaning tasks can be found in both public areas and in industry. Figure 1.13 shows an automatically operating cleaning robot for wet cleaning the floor in a supermarket; it could of course also be used in a warehouse or production hall. In both application environments, the requirements for operation are quite similar: in the shortest possible time, "everywhere", i.e. every accessible floor surface, should be cleaned without (unintentionally) driving over sections several times. It goes without saying that the machine must not press, touch or even injure any person. In public spaces, however, considerably more disturbing influences must be expected than in industrial environments, since, for example, it is to be expected that children and young people with an affinity for technology will want to explore the limits of the system and try it out: What happens if I put myself or a shopping trolley in the way, "push" the device into a corner, press one of the many buttons, etc.

And this will probably not only happen once when the device is new and first discovered in action (as is common in the industrial environment), but again and again with other persons/customers. Even if the cleaning robot masters all these challenges, at least a reduced performance, measured in cleaned area per time, can be expected compared to industrial use. We can thus recognise a decisive difference between the use of an AGV in an intralogistics application and an STS in the public sector: The AGV has to deliver a predictable and reliable performance, so the specified execution time must be precisely adhered to for each individual transport. The—perhaps autonomously acting—STS cannot offer this reliability.

Fig. 1.14 Picking robot TORU. (Source: Magazino)

But there are also new fields of application in purely industrial environments and thus new markets for automatic vehicles. Examples include mobile robots for order picking and the Munich Start-Up Magazino (see also Table 1.2), which represents several suppliers. The mobile robot "TORU" does and can do what the logistician understands by the picking principle "person-to-goods": It drives along shelves in the warehouse, stops at the right place, picks up the ordered goods at the right level and places them either on a shelf carried along or in a container on a "normal" AGV driving behind it. A 3D image processing system ensures precise navigation in the shelf aisles and exact gripping of the—currently still cuboid—articles. It goes without saying that this mobile robot works safely, because it is now successfully used in several customer installations, i.e. simultaneously and in the same warehouse aisle, with human order pickers (Fig. 1.14).

A final example actually describes a vintage AGV application, but due to technical restrictions it has so far only represented a very small niche: There have been AGVs in outdoor use for many years, the so-called outdoor AGVs—but only in very small numbers. The reasons for this are mainly to be found in the problem area of safety technology, more precisely in non-tactile safety sensor technology, in order to allow the high speeds (= high transport performance) required in outdoor use. In this environment, which is very demanding for several reasons (weather influences, road with fluctuating surface quality, high driving speed, etc.), even with further road users in the immediate vicinity/on the same road, a number of projects are currently being developed, as certified (Performance Level d) safety laser scanners have been available for outdoor use since the end of 2018. Applications here are mainly the transport of large and/or heavy goods; quite often the load pick-up and load delivery is not done automatically but manually, i.e. the vehicles are loaded and unloaded by means of a man-operated forklift or crane.

1.4.2 New Functions and Technologies

The new functionalities and thus the new applications have been made possible by developments and innovations in various fields of technology:

Table 1.3 Functions of autonomous passenger cars, which are also required for AGV

Designation	Description
Image recognition	Objects (e.g. traffic signs), people, categories, situations—even under unfavourable optical conditions (backlighting/dazzle, darkness)
Video recognition	Interpretation of moving images, sequences and gestures—even under unfavourable optical conditions (back light/glaring, darkness)
Consideration for road users	Assessing speeds, directions and intentions—also of "difficult" road users (cyclists, children, drunk person(s), etc.)
Sounds	Recognition and localization

- More suppliers and more and cheaper products for the precise measurement/acquisition of environmental features (laser scanners, 2D and 3D cameras, radar)
- Significantly improved methods for contour navigation using the above-mentioned sensor systems, the so-called multi-sensor fusion
- Sharply reduced prices for high-precision differential GPS receivers, free RTK GPS signals available almost everywhere (in Germany)
- Lithium-based batteries in combination with contactless (inductive) charging
- Sensor systems and software from the automotive sector: driver assistance systems

We would like to go into the latter aspect in more detail: Today, modern passenger cars are already able to move largely automatically, even at high speeds. Of course, there is no doubt that car developers have the advantage of being able to design functions for safety without having to be responsible for safety. After all, the car driver is currently still responsible for safety—and this will probably remain so for some time to come. Irrespective of when the first autonomous passenger cars, i.e. cars that under all circumstances operate safely without a driver, will be approved for road use, there are, however, a number of requirements for these passenger cars that must also be met by future STS; these are summarised in Table 1.3.

Summary and Outlook The STS is smaller, more manoeuvrable and more intelligent than the vintage AGV, i.e. the STS can be used in addition to the existing "large" AGV solutions to complete logistics automation—not only indoors but also outdoors.

For all applications, the following applies: the intrinsic intelligence of vehicles, one could also call it the degree of autonomy, will increase and STS/AGV will benefit from the developments in driver assistance systems and autonomous passenger cars.

The AGV was and is a fascinating technology. "Robot vehicles", which move the material as if guided by a ghostly hand, always have an emotional side in addition to the purely technical and economic aspects. Even though we humans are becoming increasingly accustomed to automated systems, the AGV of the fourth epoch will become much more intelligent and powerful than those of the previous epoch, so that the fascination of automated guided vehicles will remain in future—the world will become more colourful!

Technological Standards

If one resumes the division of AGV history into epochs, AGV development is now in its fourth epoch. The third epoch has left its mark on AGV technology: a stable technology standard has developed with which customer-specific, reliable system solutions can be realised. *It is important to note that this standard will continue to exist and be justified in the fourth AGV epoch and will not be replaced, but merely expanded!* This chapter is about the description of this technology standard.

The aforementioned VDI Expert Committee 309—Automated Guided Vehicles developed a definition of an AGV in the 1990s and published it in VDI Guideline 2510:

> Automated Guided Vehicles (AGVs) are floor-supported, self-propelled means of transport which are controlled automatically and guided by a non-contact guidance system. They are used for materials transport, i.e. for pulling and/or carrying of goods to be conveyed, by means of active or passive load handling devices. This guideline deals with wheel driven vehicles. Rail-guided vehicles, air-cushion vehicles, and walking machines are excluded from the scope of the guideline.

The same directive also contains a definition of AGV Systems:

> Automated Guided Vehicle Systems (AGVS) are inhouse, floor-supported material handling systems comprising automatically controlled vehicles whose primary task is materials transport rather than the transport of passengers. They are used inside and outside of buildings and essentially consist of
>
> - one or several automated guided vehicles,
> - a guidance control system,[1]

---

[1] In this guide we will use the term *AGV master control system* instead of *guidance control system*; the term *guidance* will be used in the context/meaning of *track guidance*.

© Springer Fachmedien Wiesbaden GmbH, part of Springer Nature 2023
G. Ullrich, T. Albrecht, *Automated Guided Vehicle Systems*,
https://doi.org/10.1007/978-3-658-35387-2_2

- devices for position determination and localisation,
- data transmission equipment,
- and infrastructure and peripherals.

In the current situation, these definitions no longer seem entirely appropriate and in need of revision for various reasons: there are new providers on the AGV market almost daily. This is accompanied by an inflation of terms. In particular, the automated vehicle, previously abbreviated as AGV, is receiving new names, such as autonomous robot and other inventive word creations. And the term "autonomous" also seems inevitable with many—especially new—players in the market. However, the characteristics of the vehicles are often overrated to such an extent that it is hardly possible to speak of an autonomous vehicle. Moreover, it remains to be clarified whether autonomous behaviour of automated vehicles brings advantages in every case or perhaps leads more to chaos. Finally, as already described in Sect. 1.4, some things have changed with regard to the possible uses of AGVs, for example the tasks of the system, which today very much go beyond pure transport and also extend to services.

The definition of AGV system was therefore adapted[2] by the VDI technical committee and will be published in the revised version of VDI guideline 2510 in 2022. The definition of an AGV was dispensed with because there are now so many different names for automated vehicles that it is no longer possible to establish a regulated designation.

> Automated guided vehicles (AGVs) are floor-supported systems that are used inside and/or outside of buildings. They essentially consist of one or several automatically controlled vehicles which are guided by a non-contact guidance system and which are equipped with their own traction drive and, if required, consist of
>
> - an AGV master control system,
> - devices for position determination and localisation,
> - data transmission equipment,
> - and infrastructure and peripheral devices.
>
> The essential task of an AGV is the automated transport of materials. In a broader sense, AGVs also include systems that are used for service tasks such as handling, monitoring, cleaning, mobile information and guidance—also in areas accessible to the public.
>   Excluded from this and not considered in the following are devices that are made available on the market as consumer products in accordance with the Product Safety Act.

The structure of this chapter may seem illogical at first glance. Here we do not look at the AGV system in a strictly system-hierarchical way, but instead focus on where the automation idea determines the technology, i.e. on the driverless aspects. These result from the functional differences to driver-operated vehicles, such as classic forklift trucks:

---

[2]Ullrich, G.: 30 Jahre Faszination FTS. Logistik für Unternehmen 06-2017, pp. 38–39, Springer-VDI-Verlag Düsseldorf.

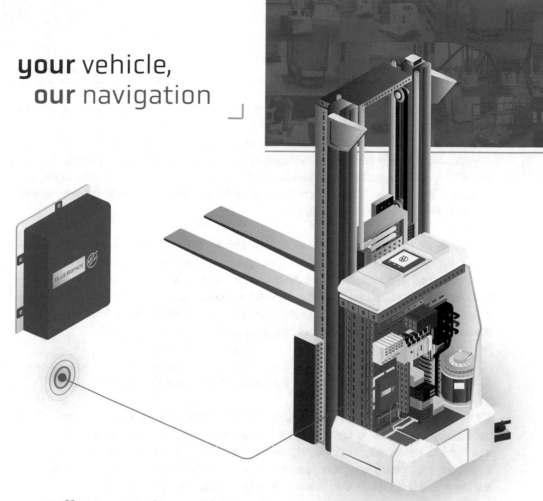

- Automated vehicles localize themselves in a known environment without being directly controlled by an operator.
- Automated vehicles guarantee their safe operation, i.e. they take care of any personal safety as well as protection against damage to the load and surrounding equipment.
- AGV systems organise themselves in terms of optimal processing of transport orders.
- Automated vehicles integrate into existing environments and are able to communicate on-demand with adjacent systems.

Therefore, we start with navigation and safety, ultimately the two most important functions of an AGV. Then we look at the AGV master control system, because it is even more important for the organisation of an AGV system than the individual vehicle. We then deal with the vehicle and its main components in the third section. The fourth section deals— last but not least—with the stationary environment of the AGV, which includes the infrastructure and peripheral devices and equipment.

2.1   Navigation and Safety as Central System Functions

From our human activity, we understand the two functions of "navigating" and "safety" not necessarily as separate, but as integrated. When we walk or run, we try to stay on the paths and reach the destination (navigation). At the same time, and always and continuously, we take care not to collide with anyone or run into anything (safety). So we practise safe navigation, and we do so always and with all our senses—if all goes well.

The AGV in its third epoch could not yet do that. There are still quite different functions performed with different techniques and components. The AGV follows a physical or virtual guiding line until a separate safety system orders it to stop. Certain navigation components and control parts act until, for example, a safety scanner and its emergency stop circuits respond.

The fact that these two functions will be integrated in the future is a current topic of the fourth AGV epoch that has already begun. Here we use the actual situation for a simplified sequential description of the functionalities, which also corresponds to the current technical standard.

2.1.1   Navigation

According to DIN, *navigation*[3] is understood to be measures for vehicle guidance, with the help of which it is determined,

---

[3]DIN 13312:2005-02 "Navigation—Terms, abbreviations, formula symbols, graphic symbols".

(a) where the vehicle is located (localisation),
(b) where the vehicle would end up if no action is taken to alter its movement, and
(c) what needs to be done to safely achieve a desired goal, if necessary along a given path.

If we put ourselves for a brief moment in the role of a forklift truck driver who has to pick up a pallet in the warehouse, for example, and then bring it to the goods out area, we realise that the driver—possibly unconsciously—is constantly navigating: at all times he knows where he is and he knows the (shortest/best) way to the destination—provided he is trained and knows the location. And of course he also knows where to accelerate, brake and steer. In an automated vehicle, this human driver does not exist—his tasks must now be carried out by a computer with software and various sensors and actuators. In order for this to succeed and lead to a comparable result, an AGV or the vehicle computer software must also navigate, i.e. it must constantly determine the current vehicle position, compare it with the desired path and, if necessary, make small (steering) corrections, and accelerate, brake and steer at the right places. The fact that this task is technically challenging but fundamentally solvable is probably obvious to everyone—after all, it was already possible to fly to the moon by computer in the 1960s. But achieving the goal with economically justifiable effort, which usually requires the use of components available on the market, is yet another challenge. Therefore, over the years, technically very different navigation procedures for AGVs have been developed, some of which differ significantly in the components used, their costs and the technical possibilities.

It should be mentioned at this point that we do not want to understand the term *navigation* narrowly in the sense of the aforementioned definition, but rather as representative of the various track guidance and localisation methods of AGVs—which is more in line with the common usage of that term in industry.

Back to the theory, to some basic connections and important technical terms:

The AGV moves in a fixed coordinate system whose base area corresponds to the driving area of the AGV (e.g. a warehouse). On the vehicle itself, a vehicle coordinate system can be spanned, the origin of which is usually located in the centre of gravity of the base area or in the centre of one of the vehicle axes. Within this mobile coordinate system, it is not the vehicle movements that are described, but movements relative to the vehicle, e.g. load movements, or also the movements of the driving and steering motors.

The fixed coordinate system—engineers and surveyors call this the "world coordinate system"—is usually a Cartesian coordinate system and will usually have its origin in a corner of the hall or at the farthest corner of the area of operation. The AGV will then move exclusively in the base of this coordinate system. Movements in the fixed height axis do not actually occur, except for movements of the AGV in a lift/elevator from one level to the next, where the specification of a level number is sufficient to describe the "height" (Fig. 2.1).

It is conceivable that halls or parts of halls in which AGVs move around are at different height levels and are connected to each other by ramps. This will also not (have to) be taken into account in the fixed coordinate system, as an AGV always drives on the ground and

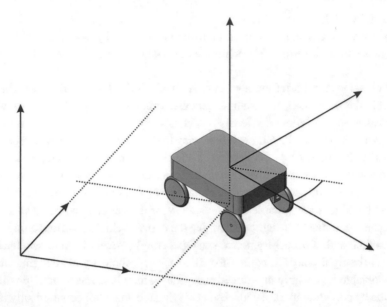

Fig. 2.1 The AGV in the world coordinate system and with its own coordinate system

knowledge of the absolute height, e.g. above sea level, or the relative height compared to the origin of the coordinate system is not relevant for the driving manoeuvres.

Important for the processes is now the determination of the position of the vehicle in the world coordinate system. This is usually described by the two translatory coordinates of the ground plane and a rotatory coordinate, i.e. the orientation in the ground plane. Now two fundamental processes take place, which are more or less pronounced depending on the method: the dead reckoning and the bearing.

Dead reckoning—also called odometry in the case of land vehicles—refers to the determination of position by means of internal sensors for measuring the direction of travel as well as the distance travelled or speed, starting from a known starting position. Dead reckoning has its origins in seafaring and has been used there for centuries. In the case of a ship, it is carried out with the help of a compass, clock or log[4]; in the case of an AGV, angle encoders and incremental encoders on the wheels, time counters and, if necessary, a rotation rate sensor (gyroscope) are used.

Odometry is inherently error-prone, e.g. due to wheel slippage, changes in wheel circumferences due to wear and changing load weights, as well as difficult calibration of the directional stability. Here, AGV manufacturers often go to enormous lengths to increase accuracy and reliability. For example, special measuring wheels (one or two) are integrated into the chassis, which register the movement of the vehicle as accurately as possible, independent of drive or steering influences. In addition, in order to be able to

---

[4]Device for measuring the speed of ships, measures the speed through the water.

record as accurately as possible the rotation of the vehicle around its vertical axis—while driving along curves, but also possible during a straight-ahead manoeuvre as a result of an incorrect adjustment of the straight-ahead movement—a gyroscope is also installed. These systems all improve the accuracy, but the errors eventually add up to unacceptably large values despite all the effort. Therefore, timely and regular bearings are necessary.

These bearings use either stationary passive markers or even active technologies. Fixed marks can be artificial: On a ship, one will survey a lighthouse, fairway buoy or prominent building on land with a compass, the AGV will survey marks placed in or on the floor, or reflex marks placed on pillars or walls. However, fixed marks can also be natural: The skipper on the sailboat may recognise headlands or other prominent features on land, which he or she can aim at with the compass and enter as bearing lines on the nautical chart. The AGV may be able to recognise certain building contours and use them to determine its position. The terms artificial and natural are both commonly used, but probably not very well chosen: What is natural about a wall or a building contour? It is probably more accurate to characterise the markers used for navigation as *reference markers* (for artificial markers) and *environmental features* (for natural markers).

The best-known modern representative of an active technology for localisation is the GPS[5] system. Here, the GPS receiver uses the radio signals emitted by the GPS satellites to calculate its current position in a world coordinate system (e.g. WGS84). The AGV uses these measurements to determine its own position—similar to the navigation device in a car. Another, much older active system for direction finding uses RFID transponders (usually in the ground), which only become active when they receive an appropriate signal from an antenna mounted on the AGV, to which they then respond.

Finally, the term location should be explained: Location is a sometimes different description of position. While position means an exact representation of coordinates, location can contain other information, e.g. when certain actions are required. These can be: branches, load transfers, docking points, stations or points in the layout that require separate flashing or warning signals as well as a change of driving speed. Thus, individual, specific positions in the layout can be defined as locations and perhaps numbered consecutively.

We will now turn to the navigation procedures. We will not go into the theory of the methods in too much detail, but concentrate on the benefits for the user. Finally, Sect. 2.1.1.6 compares the relevant methods with their characteristic properties in a table.

2.1.1.1 The Physical Track Guidance
Automated guided vehicles that navigate or drive on physical guiding lines (Fig. 2.2a) use devices on or in the floor. The most common variants are:

---

[5] For details about the GPS technique please refer to Sect. 2.1.1.5 Radio Localisation.

- The *active-inductive track*, in which current-carrying conductors are embedded in the ground. An alternating current (4–20 kHz, typically 100 mA) is driven through the conductor loop by a so-called frequency generator. The current-carrying conductor is surrounded by an alternating magnetic field that is not influenced by dirt, rain, snow or the road surface (concrete, asphalt). Only a certain distance to metallic covers and installations (e.g. steel mats for reinforcement) must be maintained. The alternating magnetic field is evaluated by an "antenna" which is mounted under the vehicle. Two coils are arranged in the antenna in such a way that the alternating magnetic field can induce a voltage in each coil, the level of which depends on the lateral distance to the conductor. If the conductor is exactly below the antenna's center, the two induced voltages are equal; if the conductor deviates to the left or right, one of the two voltages is correspondingly higher or lower. This differential voltage is then used to control the steering motor. The permitted distance between the antenna and the conductor, which roughly corresponds to the ground clearance of the vehicle, is 30–70 mm, depending on the sensor used and the current fed into the conductor loop (Fig. 2.3). Inductive track guidance became the most important technology of the second AGV era. In modern systems, however, it hardly exists any more in the original form described here.
- The *magnetic guidance track*, in which a metal strip approx. 5–10 cm wide is bonded to the floor. A sensor consisting of two to three magnetic field sensors underneath the vehicle detects the field change at the edges of the metal strip. The sensor output signal thus again corresponds to the lateral deviation of the sensor from the centre of the metal strip and can be used to control the steering motor. There are also methods in which magnetic strips are laid instead of simple metal strips. The reading distance is typically 30–50 mm.
- The *optical guidance track*, where a colour line with a clear colour contrast to the surrounding ground is either painted or applied with a special self-adhesive colour tape. A suitable camera sensor system under the vehicle + evaluation software—either directly in the camera or in an external evaluation computer—also uses edge detection algorithms and thus calculates the lateral deviation from the centre of the track and in turn generates control signals for the steering motor. Today's digital signal processing also allows the detection of heavily damaged tracks, making this method quite robust— but it is certainly not intended or suitable for working environments with a lot of dirt and grease on the floor. In addition, the cameras can evaluate codes applied next to the track in the form of barcode or QR code labels for position determination. The reading distance (= ground clearance) is usually between 30 and 70 mm (Fig. 2.4).

If the layout is simple—i.e. without many branches—the magnetic or optical track is more suitable for cost reasons. Unless, of course, contactless energy transmission is used, which works with a double conductor laid in the ground and, in addition to energy, also provides navigation as a by-product. This technology is discussed in more detail in Sect. 2.3.4 "Energy supply" (Fig. 2.5).

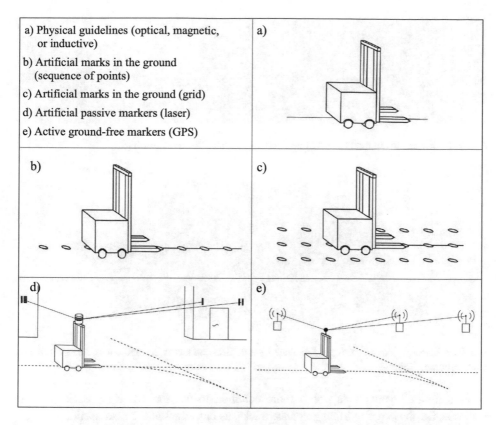

Fig. 2.2 Common guidance and navigation methods: (**a**) the rigid methods with physical guidance line and (**b–e**) the "free" (non-rigid) methods with virtual guidance line. (**a**) physical guidance lines (optical, magnetic or inductive). (**b**) **artificial** passive or active marks in the ground (sequence of points). (**c**) artificial passive or active marks in the ground (grid). (**d**) Artificial passive ground-free marks (retro-reflector tags). (**e**) Active ground-free marks (GPS)

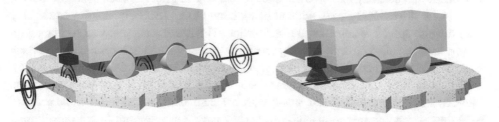

Fig. 2.3 Principle sketch for inductive and optical track guidance. (Source: Götting)

Metal or coloured strips are mostly used in simple layouts as well as in low-end AGVs. Typical applications are found where AGVs are used to link assembly workstations or vehicles act as so-called "rolling workbenches".

Fig. 2.4 Devices for inductive and optical track guidance. (Source: Götting)

Fig. 2.5 Principle sketch for track guidance by a double conductor in contactless energy transmission. (Source: Götting)

In complex layouts today, it is more common to use the non-rigid track guidance methods, i.e. those with virtual guidance tracks, as described in the next section.

2.1.1.2 Navigation with Support Points

If, for cost reasons as well as to increase the flexibility of the drive track design, one would like to do without the continuous guidance line described in the previous section, one can use the so-called base point or grid navigation. In this case, artificial markers are arranged in or on the ground along the route in a more or less regular grid. Between these grid points, the vehicles drive "freely", i.e. without being bound to a physically existing track. One therefore also speaks of so-called free navigation. The AGVs thus use dead reckoning on the one hand, but also direction finding (bearing) to determine their location. The markers used for this can be permanent magnets, RFID transponders or QR code labels (Fig. 2.6).

The inexpensive permanent magnets are mostly made of neodymium-iron-boron (NdFeB) and have a cylindrical shape with a length of 5–30 mm and a diameter of 8–20 mm, depending on the manufacturer. They have a simple north/south polarity and are exceptionally strong at 1100–1250 mT.[6] They are laid in specially drilled holes in which the magnet is fixed with the help of epoxy glue. The floor can then be resealed with a

---

[6]Tesla (T)—Unit of magnetic field strength.

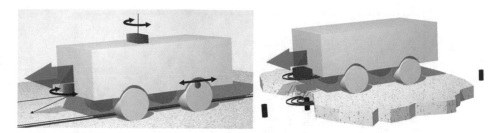

Fig. 2.6 Principle sketch for dead reckoning (left) and for magnet or transponder navigation (right). (Source: Götting)

layer of epoxy or vinyl, which protects the magnets against mechanical influences and damage.

The magnets can be laid as a grid or as a linear sequence of points. The area-covering grid allows for greater layout flexibility; when laid as a point sequence, far fewer magnets are required due to the principle involved. The point or grid spacing is determined by the accuracy requirements for the driving movements, the vehicle kinematics and the vehicle dimensions. The achievable quality of dead reckoning also has an influence, which can be improved by using a rotation rate sensor.

The number and position of the drilled holes/magnets are determined by the AGV manufacturer. Magnets in a sequence of points usually have a spacing of 1–10 m. They are laid in a grid pattern at a distance of less than the width of the vehicle, whereby usually only every second magnet is placed on each laying line for reasons of effort and cost and the lines are staggered against each other (Fig. 2.7).

Figure 2.8 shows a magnetic sensor bar (MSB) that was developed specifically for use in AGVs. Hall sensors are used to measure the magnetic field strength of the floor magnets. These convert the magnetic field flowing through them into a voltage that is proportional to the field strength.

The length of an MSB is configurable. The bar consists of one or more groups of eight hall sensors each. Its maximum length is limited to the width of the vehicle because it is mounted across the lane under the vehicle.

Reading distances, i.e. the distance between the MSB and a magnet, from 10 to 60 mm provide a measuring accuracy of less than ±2 mm.

Another product on the market is the MMS (Magnet Measurement Sensor), which is designed for both indoor and outdoor use. It does not achieve the very high accuracy of the MSB, but only approx. ±5 mm. On the other hand, it achieves this at high reading distances (up to 200 mm) and at crossing speeds of up to 80 km/h. It is therefore not only intended for classic AGV use, but also for fast-moving semi-automated vehicles in outdoor use, such as public buses, where the driver receives assistance when approaching stops (Fig. 2.9).

In outdoor areas, quasi-active transponders are often embedded in the ground instead of passive magnets. These are supplied with energy by the reader unit under the vehicle via induction, which they then use to send their identification (code ID) to the reader unit. At

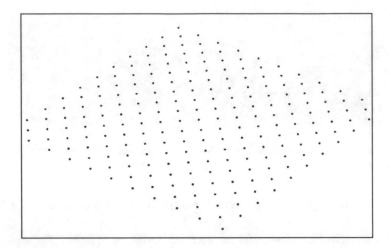

Fig. 2.7 A staggered magnetic grid

Fig. 2.8 A Magnetic Sensor Bar for indoor use; standard length: 387 mm, height: 43 mm, width: 50 mm. (Source: MLR)

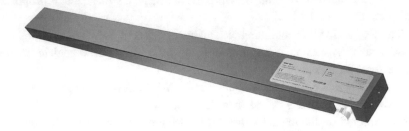

Fig. 2.9 A Magnet Measurement Sensor (MMS) for indoor and outdoor use; lengths from 530 to 2210 mm, height: 30 mm, width: 60 mm. (Source: Oceaneering)

the same time, antennas in the reader unit ensure that the position of the transponder relative to the antenna is precisely measured (Fig. 2.10).

In addition to the absolute coding and the possibility of storing additional layout information for location determination, the advantage over magnetic navigation is the

Fig. 2.10 Reading units for transponders. (Source: Götting)

greater reading distance, which provides more ground clearance for the vehicles. However, the devices are significantly more expensive and larger than a magnetic sensor bar.

2.1.1.3 Laser Navigation

Laser navigation—more precisely: laser triangulation, laser localisation—is the most prominent and currently most widespread representative of free navigation. Markers made of retroreflective material are mounted on walls, pillars or machines, usually at a height above the heads of the employees. They are accurately measured by a rotating laser scanner mounted on the vehicle at the same height, even over longer distances (Fig. 2.11).

Depending on the measuring method of the sensor (with or without distance measurement), at least two or three marks must be visible at the same time to determine the position. To avoid ambiguities, it may be helpful to code the marks. Based on the sensor readings, the evaluation software then calculates the current resulting absolute position (x, y, angle α) of the vehicle in the fixed coordinate system of the hall in real time, i.e. while the AGV is moving. Depending on the sensor and provider used, this absolute position is available five to ten times per second for further processing by the vehicle computer.

The procedure offers a high degree of flexibility with regard to the drive track design, as the positions of the reflector marks have no relation to the driving path of the vehicles. The drive track for the vehicles "only" exists as software, i.e. purely virtually, and can therefore be changed as often and easily as desired without having to change anything about the reflector marks. New drive tracks can either be programmed offline, i.e. without a vehicle but using a software tool on a laptop, or by means of a learning drive using a vehicle (the so-called teach-in method).

The choice between the two methods—grid navigation or laser triangulation—is made not least on the basis of the following criteria: ground magnets require (minor) work on the ground; a clear view of a sufficient number of reflector marks must be possible from any layout position; and it must be possible to mount the laser sensor at a sufficient height on the vehicle.

Fig. 2.11 Principle sketch (left) and laser scanner including reflector (right) for laser navigation with retro-reflectors on walls and columns. (Source: Götting)

2.1.1.4 Contour Navigation

In addition to or even instead of artificial bearing marks, "natural" bearing marks (landmarks such as walls, pillars, door niches, etc.) can be used to guide vehicles. It is important that these marks are clearly visible and that their position is unchangeable.

Mounted on the AGV, a laser scanner contactlessly scans its surroundings and continuously measures the positions of the fixed landmarks while the vehicle is moving or standing still. By comparing the coordinates of these landmarks with the coordinates previously stored in a map in the vehicle computer during commissioning and configuration, the evaluation software then recognises valid landmarks along the roadways. These are used to determine the own current position and orientation on the drive track and are made available to the vehicle computer. The vehicle computer continuously compares this current position with the target position and corrects the course deviations of the vehicle, which are caused by tolerances in the vehicle geometry, different loads/load weights, wheel wear etc. (Figs. 2.12 and 2.13).

Since it is widely used to use optical sensors for obstacle detection, it makes sense to also use these sensors to guide the vehicle through natural (= already existing, not additionally mounted) bearing marks. A typical application is the use of a distance-measuring safety laser scanner to drive parallel to a wall. In this case, the wall only serves as an auxiliary orientation; a complete position determination/navigation is only possible when additional procedures, such as edge detection or, if required, artificial bearing marks, are added.

Although the use of a safety laser scanner to measure the environment is obvious—for cost reasons, since an additional sensor can be saved—it has the disadvantage, which should not be underestimated, that at the height at which such scanners are mounted (scanning plane approx. 10–15 cm parallel to or slightly inclined towards the floor), the environment is subject to relatively strong changes: along the roadways, all kinds of objects are often placed for more or less long periods of time, which can make the comparison between the stored map data and the environment currently perceived by the sensor considerably more difficult. In addition, until around the end of 2017—when the

Fig. 2.12 Laser scanner for determining the position using reflectors (artificial landmarks) and environmental contours (landmarks); left: Functional principle, right: NAV350. (Source: SICK)

microScan3 from the company SICK came onto the market—the angular and distance resolution of the safety laser scanners was not suitable or not sufficiently accurate, at least for navigation or AGV applications that required a high precision of the driving and positioning processes. If, on the other hand, the navigation laser scanner—as an additional and precisely measuring sensor—can be mounted on top of the AGV at a height of 2 m or more, the environment that can be evaluated at this height is significantly less volatile, and the results are correspondingly more stable and accurate. However, there is a vehicle type that has been in high demand and frequently used in the recent past, the underride AGV (also known as "turtle", see also Sect. 2.3.1.5), for which only mounting the sensor close to the ground is possible, so that the disadvantages mentioned above must be accepted.

If the laser scanner is additionally swivelled around another axis, it is possible to create a 3D image of the surroundings. In this way, however, considerably more data is generated than when using a static 2D scanner, i.e. the evaluation of the data requires more powerful hardware and software and ultimately still more time than the evaluation of two-dimensional images and is therefore in principle rather unsuitable for fast-moving vehicles. The advantage is that it is now possible to locate oneself in buildings by the ceiling (ceiling navigation). The view of the ceiling is usually free of obstacles and can therefore be used reliably (Fig. 2.14).

Current development and project findings show that a positioning accuracy in the range of about ±20 mm can be achieved by exclusively using the ceiling navigation. As a by-product of the navigation, the overall concept offers obstacle detection for the vehicle

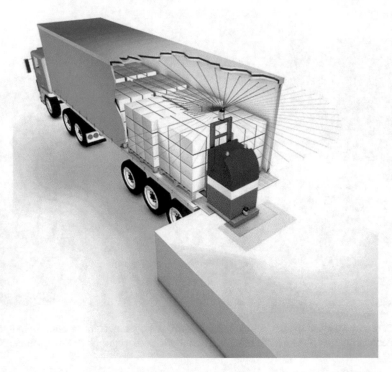

Fig. 2.13 Positioning by contour measurement using the example of a truck load. (Source: SICK)

including the load by means of the swivelling scanner. Although this is not "safe" obstacle detection in the sense of personal safety and protection, static obstacles, such as the hand control unit hanging in the drive track of the AGV or the hook of a gantry crane, are detected and a braking process is initiated as a result. Overhead cranes, by the way, are not an obstacle for this technology: although they temporarily or partially prevent a clear view of the ceiling, the software is able to hide the height section in which the crane is moving.

A compromise between static and rotating 2D laser scanners is the use of a multi-layer scanner: Here, 2D measurement values of the environment are also generated—but not only in one measurement layer, but in several. The devices available on the market offer four measurement planes in the simplest (cheapest) case, high-end devices currently offer up to 128 measurement planes (Fig. 2.15). In reference to the term "RADAR" (Radio Detection and Radiation), this type of device or the functional principle is called "LiDAR" (Light Detection and Ranging).

Furthermore, it is possible and common to mount one or more cameras on the AGV instead of the aforementioned laser scanners in order to record and measure the environment and to store the recorded data, which is processed by software, in a 3D map image of the environment.

Finally, the latest type of environmental navigation introduced to the market in 2019 will be presented, which is also based on the evaluation of camera images—namely high-

Fig. 2.14 The principle of the swivelled laser scanner mounted on the roof of a forklift truck, which thus becomes an AGV with free navigation. Above, searching for natural marks on the ceiling: the red, pivoted scanner; below, at shin height: the yellow, permanently installed personal protection scanner (*Ceiling Navigation* by company Autonomous Navigation System (ANS), Source: Siemens)

Fig. 2.15 Multi-layer laser scanners, MRS1000 and LMS1000 on the left, VLP-16 Puck and AlphaPuck-VLS128 on the right. (Sources: Sick and Velodyne)

resolution images of the floor underneath the AGV. At LogiMAT 2019, the Dutch company ACCERION presented the "Jupiter" sensor system, a positioning system that delivers odometry data (change in sensor or vehicle position in longitudinal and transverse direction as well as change in orientation) 100 times per second with an accuracy of up to ± 1 mm and $\pm 0.1°$ (Fig. 2.16). As these data are derived "only" from the changes in the ground images, they are of considerably better quality/accuracy than wheel-based measurement values, as they are slip-free. In addition, the system also provides absolute position data when—after a teach-in drive that was started at a starting point with position values X, Y and orientation (yaw angle) specified by the operator—the AGV or the sensor passes over and recognises any previously taught ground location again.

Fig. 2.16 Floor sensors "Jupiter" (left) and "Triton" (right). (Source: ACCERION)

With the "Triton" device, another sensor is available and, with significantly smaller dimensions, is aimed in particular at applications in small AGVs. With a higher measuring rate (150/s), but the same accuracy as the larger "Jupiter", it will only provide absolute position values, i.e. the relative dead reckoning data must then be generated as before on the basis of other sensor technology.

2.1.1.5 Radio Localisation

Let us now turn to the active technologies with artificial ground-free markers. In order to localize oneself in very large spaces or even in the open field, passive markers are usually not sufficient. Active transmitting markers, e.g. lighthouses in maritime navigation, have the advantage that their bearing direction in relation to the vehicle can be clearly determined even over very large distances. When using several of these marks, the location and direction of the vehicle can be determined.

More important, however, are procedures in which self-localisation is carried out, e.g. by means of time-of-flight measurement to satellites (GPS—Global Positioning System) or stationary, coded radar reflectors. A prerequisite for accurate and reliable position determination is that there is a clear line of sight between the GPS antenna on the vehicle and the satellites or radar reflectors. Therefore, satellite navigation can only be used outdoors on open ground.

General information on GPS can easily be found on the internet, which is why we do not need to go into it here. Compared to the "navigation system" that everyone is certainly familiar with from their car, however, it is important to realise that the accuracy that can be achieved by means of such (quite inexpensive) devices is by no means sufficient for the operation of an automatic vehicle. Here, (repeatable) positioning accuracies in the single-digit centimetre range and angular accuracies of about $0.1°$ are usually required. For such an accuracy, the so-called "Real Time Kinematic dGPS" is needed, which, however, also requires a free cone of sight upwards to the satellites of approx. $15°$ (Fig. 2.17). Narrow urban canyons, metal cranes and bridges severely limit the possibilities of use.

In built-up areas, e.g. between tall buildings or in large, open but roof covered halls, GPS is unsuitable. Here, only a so-called LPR (Local Positioning Radar), i.e. an "indoor

Fig. 2.17 Principle sketch for navigation using GPS (left), and GPS receiver. (Source: Götting)

GPS" can be set up and used. Instead of highly accurate, mobile and expensive satellites, relatively inexpensive radio beacons are used stationary in the area of action. Here, too, the signal's time-of-flight to the beacons is measured and thus the location of the vehicle is determined. With a favourable arrangement of the beacons, even a built-up area can be sufficiently "illuminated". However, the system is much less accurate than the complex GPS, rarely ±10 cm, mostly only ±30 cm measurement accuracy is achieved. This is usually sufficient for locating man-operated forklifts or other industrial trucks, but too inaccurate for AGV navigation.

2.1.1.6 Comparison of the Methods
The different navigation methods and their advantages and disadvantages are tabulated below (Table 2.1).

2.1.2   Safety

When a machine—and this is undoubtedly what an AGV is—is placed on the market or operated in Europe, numerous regulations must be observed that deal with the safety of such a machine. The legislation in the European Union is probably the strictest in the world. The legislator, the standard-setting institutes, the VDI (Association of German Engineers), and the German Social Accident Insurance Institutions ensure with their technical regulations that the potential hazards posed by automated guided vehicles are minimised. This is so successful that there are almost no reportable accidents caused by AGVs.

One may be of the opinion that Europe even overshoots the mark, resulting in EU products being too expensive for the world market. But these high requirements are not only implemented by the domestic manufacturers in the safety systems, but are also transferred to the overall quality of their products. Moreover, this situation is certainly partly responsible for the fact that suppliers from Asia and America have so far struggled to gain a foothold in the EU market. So our high safety requirements have advantages and disadvantages.

We want to divide the topic into four sections: First, we want to create an understanding of the legal situation, then highlight the obligations for the manufacturer and the operator, and finally take a closer look at the safety technology used.

Table 2.1 Comparison of the navigation methods

Procedure	Advantages	Disadvantages
Conducting wire (active inductive)	– Proven technology – Simple vehicle control – Complex layouts with many branches possible with multi-frequency/multi-conductor systems – Integrated safety: stop when stringline is de-energised	– Outdated technology – Inflexible – Elaborate floor installations – Layout changes extremely expensive – Susceptible to faults due to guide wire breakage
Optical or (passive) magnetic guidance track	– Inexpensive technology (low cost, "simple solution") – Simple layout is quickly put into operation – Simplest system control: stop when guidance lane is interrupted or in case of additional ground markings	– Low flexibility – Susceptible to failure due to damage to the colour coating or to the metal band
Inductive power transmission (double power cable in the floor for power transmission, but also for tracking)	– No (or only small) battery required – Well suited for simple line systems (AGV as assembly vehicle)	– Elaborate and costly installation – No complex layouts possible
Magnetic navigation in point sequence (line grid): Conducting wire is replaced by point sequence by permanent magnets or transponders	– Easier floor installation compared to line guidance – Limited flexibility: possible lateral deviation from the "magnetic track" up to approx. ± 30 cm	– Changes of drive tracks only possible with changes to the floor installations – Restrictions regarding ground clearance and ground condition (depending on the magnetic sensor bar used)
Area grid (optical or magnetic)	– Free navigation – Flexible within the grid area – Drive track layout can be adapted purely by software	– Floor must be prepared, e.g. laying the magnets – Restrictions regarding ground clearance and ground condition – Grid installation means high effort
Transponders instead of magnets	– Outdoor use possible without restrictions regarding ground clearance – Suitable for large, heavy vehicles – High process reliability due to absolutely coded screen dots	– More expensive than magnets – Elaborate and costly installation

(continued)

Table 2.1   (continued)

Procedure	Advantages	Disadvantages
"Free flight", or dead reckoning without bearing, but with gyroscope if necessary	– No fixed installations necessary	– Unreliable because free flight = blind flight – Accuracy only sufficient for short distances
Classic laser navigation (laser triangulation with reflector marks)	– No floor installation – Supplies absolute position – Enables free navigation – Simple layouts are quickly "learned" – Highly flexible within the areas equipped with reflectors – High accuracies possible through clever positioning of the reflectors (better than ±10 mm, $\pm0.1°$) – Small layout changes can be carried out by the operator himself	– Reflectors on the walls, columns, machines required – Reflector positions must be precisely measured (by a surveyor) – Laser head must be above the load to achieve free all-round visibility – Because of the high mast for the laser head, a (very) level floor is recommended – Reflectors can get dirty – Extraneous light influences can disrupt the system – Outdoor use only possible to a very limited extent
Laser navigation without artificial marks as. . . . . .building navigation or . . .ceiling navigation	– No reflectors or other artificial marks required – Building navigation systems usually use the personal protection laser scanner for navigation at the same time ($\rightarrow$ cost savings) – Ceiling navigation systems are oriented towards the unchangeable ceiling	– Increased software effort – Building navigation susceptible to changes along the driving course, only suitable for simple scenarios without much traffic – Often need to install additional prominent features along the roadway – Ceiling navigation requires additional complex (moving) sensors
navigation with ground camera /ground images	– Free of fixed installations: "infrastructure-free". – Very high relative and absolute accuracy (up to ±1 mm/$\pm0.1°$)	– Sensor costs – Ground surface must have a certain degree of visually evaluable structure – The floor must not be too dirty
Satellite navigation (GPS) better: dGPS (differential GPS) even better: RTK-dGPS (Realtime Kinematic dGPS)	– Free from fixed installations – Flexible	– Can only be used outdoors – There must be a free opening angle of at least 15° to the top. – High driving and positioning accuracy can only be realised with great technical effort

(continued)

Table 2.1 (continued)

Procedure	Advantages	Disadvantages
Radio navigation "Indoor GPS"	– Also works indoors	– High effort for installation (mounting + wiring) of the stationary radio transmitters – Achievable accuracy, even under ideal conditions, is usually not sufficient for AGV applications

2.1.2.1 Legislation

It is perhaps important to make a legal note right at the beginning of this section: no claims whatsoever can be made against the authors from the following remarks—first, because the legal situation is subject to constant change, second, because safety rules and regulations differ from country to country outside the European Union and cannot comprehensively be covered in this guide, and, third, because each AGV requires a specific solution. We refer to the "VDI Statusreport: Fahrerlose Transportsystem (FTS)—Leitfaden Sicherheit für Betreiber",[7] which is published and kept up to date by the AGV Expert Committee of the VDI association "Production and Logistics", as well as to the VDI Guideline 2510 Sheet 2 "Safety of AGVs".

Table 2.2 lists all AGV-relevant (German) laws and regulations, followed by the European standards in Table 2.3 and then the (German) guidelines in Table 2.4.

2.1.2.2 Obligations of the Manufacturer/Supplier

Manufacturers are obliged to build their vehicles in such a way that the safety requirements of the Machinery Directive are met. The AGV manufacturer must draw up so-called "original operating instructions" for his product. An instruction manual in the official language of the country of use must be supplied with each installed system—if necessary, in addition.

The AGV manufacturer must prepare technical documentation. This

- should include all plans, calculations, test records and documents relevant for compliance with the essential health and safety requirements of the Machinery Directive,
- must be retained for at least 10 years after the last day of manufacture of the AGV,
- and must be presented to the authorities upon justified request.

An obligation on the part of the manufacturer to supply the technical documentation to the purchaser (user) cannot be derived from the Machinery Directive. The machine manufacturer must confirm compliance with the applicable specifications in a legally binding

---

[7]This guideline is available free of charge on the Forum-FTS website (www.forum-fts.com; in German language only).

Table 2.2 The laws and regulations relevant for AGV systems

GPSG	Equipment and Product Safety Act Law on Technical Work Equipment and Consumer Products
9. GPSGV	Ninth Ordinance to the Equipment and Product Safety Act (Machinery Ordinance)
BGV D 27	Accident prevention regulation "Industrial trucks"
ArbSchG	Occupational Health and Safety Act Law on the implementation of occupational health and safety measures to improve the safety and health of employees at work
BetrSichV	Industrial Safety Ordinance Ordinance on Safety and Health Protection in the Provision of Work Equipment and its Use at Work, on Safety in the Operation of Installations Requiring Inspection and on the Organisation of Occupational Health and Safety at Work

Table 2.3 The AGV-relevant standards

DIN EN ISO 3691-4	Industrial trucks—Safety requirements and verification—Part 4: Automated trucks and their systems
DIN EN 954-1	Safety of machines, safety-related parts of control systems
DIN EN ISO 14121	Safety of machinery, guiding principles for risk assessment (prev. 1050)
DIN EN 1175-1	Safety of industrial trucks, electrical requirements
DIN EN 1175-2	Safety of industrial trucks, electrical requirements—Part 2: General requirements for trucks with internal combustion engines
DIN EN 1175-3	Safety of industrial trucks, electrical requirements—Part 3: Particular requirements for electric power transmission systems of internal combustion engine powered trucks
DIN EN ISO 12100-1	Safety of machinery, basic concepts, general principles for design, Part 1: Basic terminology, methodology
DIN EN ISO 12100-2	Safety of machinery, basic concepts, general principles for design, Part 2: Technical principles
DIN EN ISO 13849-1	Safety of machinery, safety-related parts of control systems, Part 1: General principles for design
DIN EN ISO 13849-2	Safety of machinery, safety-related parts of control systems, Part 2: Validation
DIN EN 1755	Safety of industrial trucks, use in Ex-proof areas
DIN EN 982	Safety of machinery—Safety requirements for fluid power systems and their components—Hydraulics
DIN EN 983	Safety of machinery—Safety requirements for fluid power systems and their components—Pneumatics

manner by issuing a declaration of conformity and marking the AGV with the CE mark. Then the AGV may be placed on the market in the European Economic Area.

The AGV manufacturer is obliged to carry out a risk assessment. For this purpose, an analysis must be carried out to identify all hazards associated with the operation of the

Table 2.4 The AGV-relevant guidelines

2004/108/EC	EMC Directive/EMC Act
	Electromagnetic compatibility (of electrical and electronic products)
VDI 2510	Automated Guided Vehicles (AGVs)
	> Implementation guideline regarding technology
	> with all parts
VDI 2510, Sheet 2	Automated Guided Vehicles (AGVs)—Safety of AGVs
VDI 2710	Holistic planning of Automated Guided Vehicles (AGVs); basics
	> Planning guideline
	> with all parts
VDI 4452	Acceptance rules for Automated Guided Vehicles (AGVs)

system. To determine the necessary measures, he must carry out the risk analysis according to ISO 14121.[8]

The risk analysis includes the steps of delimiting the system, hazard analysis and risk assessment. Then the central question is asked whether the system is sufficiently safe. If "Yes", the risk assessment is finished (positively), if "No", a "risk reduction" has to be initiated, following the 3-step method:

1. Safe design: Eliminate or minimise risks as far as possible (integrate safety into the design and construction of the machine).
2. Technical protective measures: Take the necessary protective measures against risks that cannot be eliminated by design.
3. User information about residual risks

Technical protective measures are implemented by means of protective devices (covers, fences, safety doors, safety light curtains) or monitoring units (for position, speed or standstill, etc.) which perform a safety function. Where the effect of a protective measure depends on the correct functioning of a control system, this is referred to as functional safety.

The DIN EN ISO 3691-4 (Industrial trucks—Safety requirements and verification—Part 4: Driverless industrial trucks and their systems) It takes into account the fundamental safety requirements of the Machinery Directive and the EFTA[9] regulations and serves as a basic, uniform standard. The design and technical measures are described in Sect. 2.1.2.4.

The user information includes the operating instructions and, if applicable, also all information on the proper and safe operation of the system (operator information).

---

[8] EN ISO 14121—Safety of machinery—Risk assessment, March 2008.

[9] EFTA = European Free Trade Association.

2.1.2.3 Obligations of the Operator

The operating instructions and the operator information contain the specifications for the operator. These concern the environment of the AGV as well as the vehicles.

The minimum requirements for the environment of the industrial trucks can be found in DIN EN ISO 3691-4. In particular, the following points must be observed:

Danger points are to be protected by floor markings. The installation of floor markings must be instructed by the manufacturer and carried out by the operator! The correct behaviour must be described by the manufacturer in the operating instructions. The operator must adhere to these instructions in a binding manner! No persons are allowed to stay in the marked areas!

The operator must fulfill the requirements set by the manufacturer with regard to keeping the tracks clear, clean and repaired. The details of the requirements must be described by the manufacturer in the operating instructions. The operator must adhere to these instructions in a binding manner!

When using AGVs, the operator must pay particular attention to the personnel detection systems used on the vehicle and to the load handling attachments. The operator must ensure that systems with AGVs are inspected after assembly and before initial commissioning. The purpose of the inspection is to ensure that the work equipment has been properly assembled and that it is functioning safely. For this purpose, an external expert can be appointed or one of the testing organisations operating in Germany (TÜV, Dekra, etc.) can be commissioned.

The operator must also ensure that AGVs and their attachments are inspected at intervals of no more than 1 year. These periodic inspections must cover the condition of the components and equipment, the completeness and effectiveness of the safety devices and the completeness of the test certificates.

2.1.2.4 Components and Equipment

The safety requirements for AGVs are specified in DIN EN ISO 3691-4. DIN EN 954-1— "Safety of machinery" applies to the implementation of these requirements by means of safety-related control systems. Here, five control categories are formed that describe the system behaviour in the event of failure or malfunction of the control used. The higher the category, the more serious the effects in the event of a failure—and the greater the effort required to counteract the possible failure (e.g. by using high-quality components, self-monitoring, redundancy, etc.). Table 2.5 compares the relevant functions of AGVs with the control categories assigned to them and to be fulfilled by the components used.

Let us now turn to the essential technical safety devices on the AGV. These are:

- Like every machine, the AGV has *emergency stop buttons* that must be easily recognisable and readily accessible. When pressed, the vehicle stops immediately and remains stationary until the button is unlocked.
- In order for people to be able to adequately perceive the AGV during operation, the vehicles usually have a combination of *visual* (flashing/rotating warning lights) *and*

Table 2.5 Control categories according to DIN EN 954-1

Control system		Category
Speed control	General	1
	Insofar as the stability is influenced	2
	Insofar as the effect of the person recognition system is influenced	3
Load handling	General	1
	Insofar as the stability is influenced	2
Steering	General	1
	Insofar as the stability is influenced	2
Battery charging system		1
Warning lights		1
EMERGENCY STOP		3
Personal protection system		3
Side protection		2
Bypassing the obstacle detection system		2
Stopping the driverless industrial truck in front of the obstacle		2

acoustic warning devices. For example, turns to the left or to the right are indicated by corresponding direction indicators, as in the case of a motor vehicle on a public road, and changes in the direction of travel are acoustically supported.

- The safe stop is ensured by *mechanical, self-acting brakes*. These are designed to be intrinsically safe, i.e. they require an energy supply to be released while driving. In case of an emergency, i.e. if the energy supply fails, the brakes are applied immediately (reverse principle to a car's brake). The brakes must be designed in such a way that they can bring the AGV to a safe stop even under maximum load and even with maximum longitudinal inclination of the track (downhill slope).
- Side-mounted *safety edges* and special safety devices for load handling ensure safety during operation.
- The *personal safety and protection system* is essential. It must ensure that persons or objects on the roadways ahead of an AGV or within the envelope of the AGV and the load are safely detected. If this occurs, the vehicle must come to a safe and rapid stop before persons or objects come to harm. Mechanical systems react to contact and are designed, for example, as plastic bars or soft foam bumpers. Non-contact sensors scan the danger zone in front of the vehicle with laser, radar or ultrasound or a combination of several technologies.

In the first two AGV epochs, personal protection was realised mechanically. Metal brackets or wire mesh were used, as in Figs. 1.2 and 1.3, or then plastic brackets in the 1970s and 1980s (Fig. 2.18). The soft foam bumpers (also Fig. 2.18) are somewhat more advanced because they also respond when force is applied from above, i.e. it is impossible to enter the safety device.

Fig. 2.18 Devices for personal protection; left: Plastic bumper plus safety laser scanner (Source: Dematic); right: Soft foam bumper plus ultrasonic sensors (Source: MLR)

According to DIN EN ISO 3691-4, the mechanical stirrups or bumpers must be designed in such a way that the actuating force on a test device does not exceed 400 N in the event of contact at maximum speed and load. The cylindrically shaped test device with a diameter of 70 mm and a length/height of 400 mm is ultimately modelled on the lower leg/shin of an adult. The coefficient of friction between the wheels and the ground, the braking power and the length of the safety device then determine the maximum permissible speed of the AGV.

The plastic bumpers are actuated either by cables and mechanical switches inside the construction or by a light barrier which is interrupted when a reflector glued to the inside of the bumper leaves the beam path due to deformation of the bumper. Soft foam bumpers are bumpers made of a foam-like material with fibre optics running through them. The light transmission of the fibre optics is changed by deformation in such a way that the emergency stop circuit is interrupted.

Today, automated vehicles are mostly equipped with non-contact safety laser scanners, and very often one will find products from company Sick, as they developed these sensors in the 1990s and protected themselves against imitations with a quite comprehensive pan-European patent.

For some years now, i.e. since the expiry of patent protection, there have been other suppliers whose market shares are constantly growing. Important for the non-contact safety devices is the approval by the German Social Accident Insurance Institution for use in AGVs. Some representatives of this type are the products in Fig. 2.19.

An important distinguishing feature of the devices—apart from their obviously different sizes—is their range or the size (length) of the programmable protective fields, which ranges from 3 up to 7 m. Since the length of the protective field must correspond to the expected stopping distance of the vehicle, this dimension in turn has an influence on the vehicle's maximum permitted speed. In order to cope with the different driving situations, such as speed changes, change of driving direction, cornering or docking manoeuvres, the protective field sizes can (need to) be adjusted according to the situation (Fig. 2.20). The adjustment is done fully automatically and thus allows a much more dynamic driving style. This feature, by the way, makes the essential difference to the tactile sensors, which, due to their design, have only one fixed length, namely the maximum length needed according to the possible maximum speed. On the other hand, the non-contact safety laser scanner—

Fig. 2.19 Safety laser scanners for personal protection; top (from left to right): S3000, S300, S300 mini and microScan3 (Source: Sick); bottom (from left to right): OS32C (Source: Omron), RS-4 (Source: Leuze electronic), UAM-05LP (Source: Hokuyo)

unlike the tactile sensor—must not touch an obstacle (e.g. a person's leg), i.e. this must be ensured by suitable mounting and correct dimensioning of the protective fields.

The safety laser scanners have a safe data communication interface via which control signals can be exchanged among each other—when using several sensors on one AGV—as well as with a suitable safety controller. In conjunction with safe path measurement or speed sensors, complex monitoring functions can then also be covered (Fig. 2.21).

In addition, some units have a data interface through which the measurement data from the safety laser scanners is output in real time to provide a support function for navigation or automatic load handling (e.g. pick-up a misplaced pallet).

2.1.2.5 Mixed Operation with Persons from Outside the Company

In the previous section we explained the technical safety devices for the protection of persons in internal traffic. In the case of mixed operations with persons from outside the company, e.g. tradesmen, suppliers and the like, the existing safety equipment may reach its limits. This applies in particular to critical situations, such as those that can be triggered by a suspended load, a raised fork, scaffolding or a ladder (Fig. 2.22).

Therefore, in such cases, care must be taken to ensure that persons from outside the company are made aware of safety-relevant circumstances at the place of their use and in

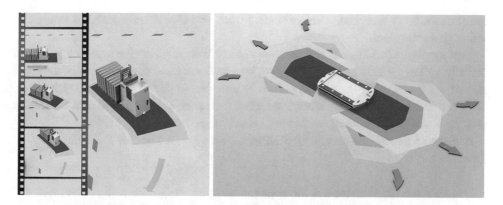

Fig. 2.20 Safety and warning fields for safeguarding the roadways of an AGV

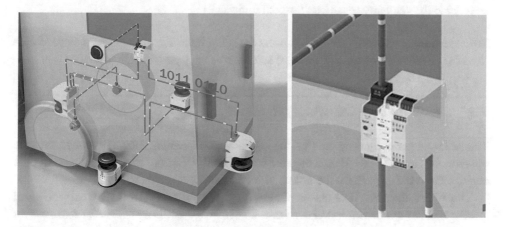

Fig. 2.21 Safe in-vehicle data communication with Flexisoft safety controller. (Source: Sick)

Fig. 2.22 Critical situations for safety equipment in the operational environment

handling the AGV technology before starting work, and care must be taken to ensure that safety shoes are worn.

Furthermore, in addition to the obligatory safety laser scanners for personal protection (usually yellow), it may make sense to provide additional sensors at the front of the AGV and, if necessary, also on the long sides, which enable 3D obstacle detection. Here, however, a distinction must always be made between sensors certified for personal protection and (non-certified) sensors for obstacle detection.

The search for suitable sensors for 3D obstacle detection has been in full swing for some time, reasons being:

• People's security awareness is increasing.
• More and more is expected from technology (here: intelligence of the vehicles).
• There is an increase in the number of cases where one has to deal with people from outside the company (example: hospital logistics).

ToF[10] cameras, radar and ultrasonic sensors are being investigated. There is a considerable need for development here and standard solutions are not yet emerging.

2.2   AGV Master Control System

The AGV master control system has the key task of integrating the AGV system into the operational environment. It also coordinates the AGVs that are in the system. This then enables the AGV to perform the tasks assigned to it.

The VDI defines an AGV master control system as follows[11]:*"An AGV master control system consists of hardware and software. The core is a computer program that runs on one or more computers. It serves to coordinate several automated guided vehicles and/or integrates the AGV system into the internal processes"*.

The AGV master control system

• integrates the AGV system into its environment (Sect. 2.2.1),
• receives and processes transport orders (Sect. 2.2.2),
• offers the operators a wide range of service options,
• and provides function blocks corresponding to the tasks (Sect. 2.2.3).

The AGV system of the third epoch is a hierarchical system. This means that the individual AGVs act intelligently, but not autonomously. The AGVs communicate little or not at all

---

[10]ToF—Time of Flight = light travel time measurement.
[11]VDI 4451-7 "Compatibility of Automated Guided Vehicle System (AGVS)—AGVS guidance control system", Status: 2005-10, VDI/Beuth-Verlag.

with each other and hardly make any decisions on their own. The actual decision-making authority lies with the higher-level AGV master control system. This means that it has the overall responsibility and the rules that are necessary to manage the entire AGV system.

Now there are plants that have no master control system at all—how does that work? On the one hand, there are systems that are very simple and do not require complex decisions. In an extreme case, think of a single towing vehicle that can only drive a prescribed distance there and back. When it arrives at its destination, it stops and waits for an employee to change trailers. The employee then presses the start button on the AGV, which then sets off on its return journey. When it reaches its destination, it stops and waits again. Such "systems" certainly have their justification, but they have a very limited functionality. There are no vehicles to coordinate, no transport orders to manage and no peripheral interfaces to operate.

On the other hand, a master control system does not necessarily have to be centralised, i.e. physically recognisable as such. Because it is not about the computer, but about functionalities that can run decentrally, i.e. "hidden" or distributed, e.g. on the vehicle computers. In our understanding, we would then still have a master control system— however, such a realisation has been uncommon until now. How this could look in the future is the subject of the fourth chapter.

2.2.1   System Architecture of an AGV System

Figures 2.23 and 2.24 show examples of two AGV systems with differences in complexity. Figure 2.23 shows a typical small system: There is a small number of AGVs with which the AGV master control system is connected via WLAN.[12] There is also a LAN,[13] via which there is a direct connection to a higher-level computer system from which the transport orders come. A VPN[14] connection for remote diagnosis is set up via the indicated telephone line.

Figure 2.24 shows the high-end configuration of an AGV system. Here one finds a multi-server system and separate operating and visualisation computers (clients). Secure data storage with an appropriate RAID[15] level is also available, as is remote communication via the internet.

---

[12] WLAN = Wireless Local Area Network, a wireless computer network in a building, also known as WiFi.

[13] LAN = Local Area Network, a computer network in a building or on a company premises.

[14] VPN = Virtual Private Network, a connection of two locally separated, independent networks.

[15] RAID = Redundant Array of Independent Disks, i.e. a "logical" drive with several independent real drives to increase data availability.

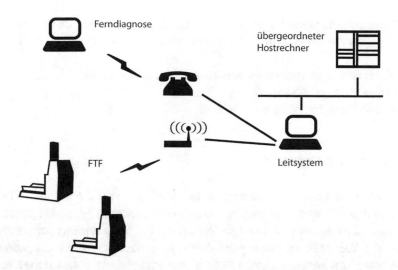

Fig. 2.23 The system architecture of a simple AGV system (according to: VDI 4451-7)

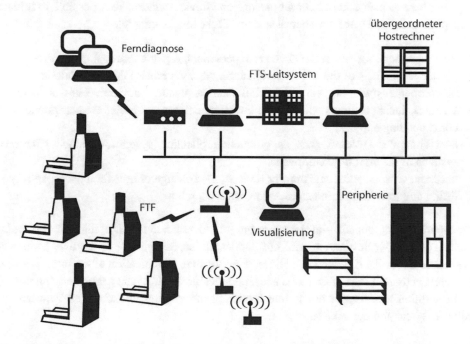

Fig. 2.24 The system architecture of a complex AGV system (according to: VDI 4451-7)

Independent wireless communication systems not only connect to the large number of vehicles, but also include picking equipment and other peripheral systems and devices. Examples include automatic doors, gates (possibly fire compartment gates) and lifts.

Data transmission to the higher-level host computers is usually carried out via local Ethernet-based networks with the TCP/IP protocol. Such host computers can be:

- Material flow control systems for production control (e.g. SAP),
- Production Planning Systems (PPS),
- Warehouse Management Systems (WMS).

2.2.2   Users and Orderers

Transport orders that are to be executed by the AGV system can reach the AGV master control system in different ways: they can be entered manually by people/employees or can be generated automatically in the broadest sense, i.e. by computers/software/sensors/ switches, etc. The AGV master control system is responsible for the execution of the transport orders. An instruction of the AGV master control system aims at the central task of an AGV system, namely the execution of transport orders. In this sense, we will group together here as user and orderer all persons and devices that make it possible to initiate transports at the AGV master control system. These are for example

- Plant operators, service and maintenance personnel, e.g. via terminal or monitor,
- service technicians of the AGV manufacturer, also via remote data transmission,
- plant operators/workers via production data acquisition devices, such as barcode scanners, indicator lights with release button or call button, tablet or smartphone with corresponding app etc.,
- host computer systems, such as production planning systems, production control systems or material flow computers,
- occupant sensors, automatic transfer stations, load change or transfer devices, conveyor lifting and lowering stations, processing stations, robots.

Transport orders contain—at least—a unique identification (ID), the pick-up point (source), the order destination (sink), if applicable the designation of the load actions at the source and at the sink, and in addition, if applicable, an indication of the order priority, the latest permitted start or end time and, if applicable, further load accompanying data.

In addition to transport orders, the following information is exchanged between the control system and the vehicles:

- Status requests and corresponding feedback from the vehicles,
- plant control or status information,
- modification of transport orders (interruption/cancellation, change of priority, etc.),
- fault messages and status messages regarding the vehicles and the layout.

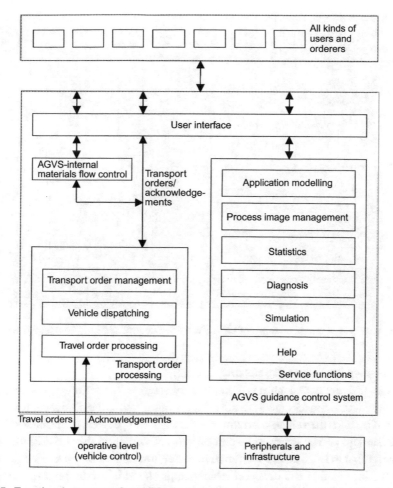

Fig. 2.25 Functional structure of an AGV master control system. (Source: VDI 4451-7)

2.2.3   Function Modules of an AGV Master Control System

Figure 2.25 shows the usual function modules of an AGV master control system.

Here we refer to the VDI Guideline "VDI 4451-7—AGVS guidance control system" and therefore only briefly describe which tasks the user interface, the transport order processing and the service functions take over.

2.2.3.1 User Interface

The user interface provides access to the AGV master control system. On the one hand, it includes the man-machine interface (system visualisation, masks and input windows, Fig. 2.26) and on the other hand various machine-machine interfaces, such as LAN and WLAN protocols.

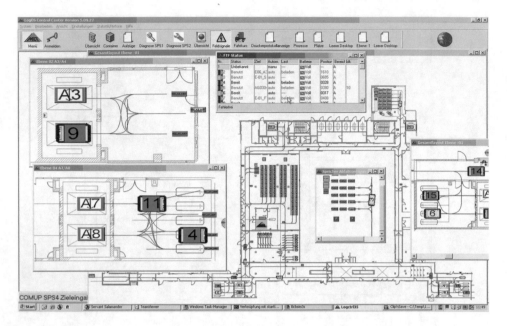

Fig. 2.26 Example of a user interface of an AGV master control system. (Source: MLR)

2.2.3.2 Transport Order Processing

The internal material flow control is a preliminary stage of transfer order processing and is only required if the transfer orders are not clearly transmitted to the master control, but must be prepared by it for the vehicles. A request such as "Need goods A at machine B" requires a conversion into one or several transport order(s) according to the classic pattern "Fetch from C and bring to D". The AGV-internal material flow control thus combines source and sink via the information and transport relationships stored in it into a transport order and sends it to the transport order management for execution. Such a rule-based generation of transport orders can take place in the following ways:

- When the AGV picks up the load, the load is identified by a scanner; a destination, i.e. the transport order sink, is then stored for this load in the internal material flow control.
- Workers enter a material request at a PDA terminal. The AGV-internal material flow control knows where to fetch this material and generates a corresponding transport order (trip to the source).
- Load transfer stations can be equipped with sensors that can be used to recognise requirements: "Empty station" = bring replenishment, "Occupied station" = pick up pallet. The necessary background knowledge for this is located in the internal material flow control, which can then generate transport orders.

The essential functions of transport order processing are:

1. Transport order management
2. Vehicle disposition
3. Drive order processing

At the end of this chain there are the *drive orders*, which are transmitted to the AGVs, carried out by them and finally reported back as completed. Ultimately, these three sub-functions contain the operational system intelligence that determines the efficiency and thus also the profitability of the system. The control system must manage the balancing act between the fastest order completion and the lowest required number of vehicles.

The **transport order management usually** receives the transport orders in the order in which they are created. In exceptional cases, the transport orders are also issued in blocks with time scheduling, e.g. from a higher-level production planning system. The transport orders are classified according to their priority or their scheduling and are constantly tested for their feasibility. This is usually given when material is available at the source and the sink is ready to receive, but it can also be subject to other criteria.

If a transport order can be executed, it is released for vehicle scheduling. If necessary, status information on the transport orders is transmitted to superordinate systems.

Vehicle scheduling determines the "most favourable" AGV for the transport orders released by transport order management. According to this decision, a transport order is transferred to transport order processing.

A wide variety of strategies can be used to determine the best ("cheapest") AGV. In the simplest case, any free AGV is commissioned. In more complex AGV Systems, however, the following criteria can also be taken into account, for example:

- The shortest (path- or time-optimised) route to the source.
- Avoidance of conceivable traffic jams and/or blockages on the drive tracks.
- The ability to pick up multiple loads at different locations.
- Forecasts of the state of the system in the near future.

Other essential functions of vehicle scheduling are the handling of empty vehicles and the battery charging strategies. When handling empty vehicles, a transport order to a parking position or a battery charging station is generated for vehicles that have processed an order and will not receive a new order in the foreseeable future.

The battery charging strategy is essentially dependent on the type of battery, the layout and the type of battery charging. Examples of battery charging strategies are:

- At the end of the shift, drive orders to battery charging or battery changing stations are generated for all vehicles.
- The state of charge of the vehicles' batteries is continuously monitored. If necessary, drive orders to battery charging or battery exchange stations are generated.

  The loading status is also monitored during loading, so that depending on the system status—e.g. number of open transport orders in relation to the number of vehicles ready to drive—a loading process can also be interrupted prematurely and a vehicle can be scheduled for transport orders again.

The **drive order processing** generates sequences of drive & action orders from the transport orders. Drive orders are linked to a destination known to the AGV, action orders can be, for example: load pick-up, load delivery, battery charging.

The drive order processing also takes over the **traffic control**. Especially at intersections and congested nodes, it ensures that the AGVs can travel without disruptions and prevents collisions and blockades. The traffic control is usually based on a division of the route into blocking areas (also called block sections), similar to the vintage procedures in rail traffic. A road user is generally only allowed to enter a blocking area if it is free of any other road user. Once a block area has been requested and allocated, it is closed to all other road users. In addition, any doors, gates, traffic lights and barriers that may be present are controlled.

2.2.3.3 Service Functions

Modern AGV master control systems offer extensive service functions. These can be categorised as follows:

- Application modelling, e.g. to change the layout in a user-friendly way or to activate or deactivate any individual AGV.
- The process image management (often also referred to as system visualisation) for informing the user about driving orders, vehicle statuses and positions, etc.
- Statistics to easily obtain meaningful data regarding the utilisation of the system (vehicles, routes, sources and sinks), number of faults, frequency of certain faults, etc.
- Diagnostics for efficient error detection as well as help functions to support the operator.
- The simulation or emulation, e.g. to analyse the system behaviour.

Application modelling creates the data basis for programming the overall system. For the commissioning of an AGV, the layout in which the AGVs are to move must be modelled as a matter of priority. This includes, among other things, the routes with driving direction information, blocking areas, load transfer points, stopping points, loading stations, etc. Depending on the track guidance method of the vehicles, other features of the layout, such as the location of reference marks, may need to be modelled. In addition to the layout, other important information is also modelled here. This includes:

- Vehicle models: describe the kinematics of the vehicle(s), attachment of sensors, technical equipment as well as logical behaviour patterns (e.g. behaviour in the event of a fault or fire).
- Peripheral models: geometric description of their shape and their position in the room as well as logical behaviour patterns, if applicable. Here, load transfer stations in particular are to be mentioned; however, all other peripheral facilities are also included, such as lifts or traffic lights.

- Simulation models: Information on the application-specific configuration of the simulation and, if applicable, the visualisation modules.

The first information flows into the application modelling in very early project phases and is continuously adapted and supplemented. More and more, this model information is stored in relational database systems, which then support consistent data management on the system's master controller.

Envelope simulation is used to graphically check the layout for collision-free movement by means of underlying vehicle models. The standard of an AGV planning is the execution of envelope curve simulations, which ensure that vehicles can move collision-free on their drive tracks. Based on the CAD layout of the operating environment, the collision-free nature of the entire route can be checked.

The **process image management** establishes, logs and informs about the temporal course of the system status. It controls and coordinates the operating modes, but also all system errors and malfunctions. It supports troubleshooting, whether on site or via remote diagnosis.

In some systems, a storage space/buffer space management is necessary. It represents a very simple warehouse management system integrated into the AGV master control system in the sense of space management. If necessary, the connection to an external warehouse management system is also possible.

Statistics functions support the analysis and optimisation of material flow events and serve to assess system utilisation. The statistics provide valuable information for the user and are important for the efficient management of the system.

Usually, a small number of basic statistics with the same structure are generated in the AGV, while other data are passed on to other computer systems for further evaluation via external statistics programmes. The data is then usually made available on the customer's own statistics servers (this can also be a simple workstation computer), so that the customer can run his evaluations himself, independently of the AGV system and the AGVS supplier.

Typical basic statistics are: Error statistics, order statistics, throughput, system/vehicle utilisation, downtimes, transport route utilisation, buffer utilisation, order throughput time, vehicle and system availability, energy balance.

Diagnostic information can also be derived from AGV error statistics. For example, if a sensor error occurs much more frequently on one AGV than on the other vehicles, this indicates a defective or incorrectly adjusted sensor.

Diagnostic systems now provide indispensable services for the safe and quick detection of problems and their elimination. Remote diagnosis offers a possibility to analyse or maintain complex systems by experts "from a distance".

The **system diagnosis** is a guided diagnosis that leads the operator through the diagnostic process in dialogue. All diagnoses to be carried out are defined, configured and activated by the operator. Based on this, automatisms perform the desired tests on the vehicle or the system and create evaluations of the test results. The operator is able to influence, activate or deactivate the diagnostic sequence. The diagnostic functions are based on error search trees, especially in the case of more "standardised" AGV control systems.

Compared to system diagnosis, **self-diagnosis** involves automatic procedures that check the system for errors and inconsistencies cyclically, on instruction or permanently. The results are stored in a suitable form and passed on to the operator. In the event of serious errors, the self-diagnosis either triggers appropriate procedures that automatically correct the problem or prompt the user to react. At the very least, the safety functions are activated that transfer the overall system or the affected components to a safe state.

Help functions "The right information in the right place at all times"—this motto applies to document and information management. Cross-process, cross-system and cross-platform, all the necessary information for the respective user can be retrieved here in multimedia form at any time. The Internet and HTML in particular have contributed significantly to this development. The following document areas can be distinguished:

- User documentation: It contains descriptions of the user interface (HMI), operating instructions and a functional description.
- System documentation: It contains information about the system installation and operation and is important for the system administrator.
- Programme documentation: It is not available for the customer; here all programs are described according to the respective documentation standard.
- Installation documents: These exist individually for each system installation and describe the specifics of the system.
- Service and maintenance documents: Information recorded by the service technicians about faults, repairs and maintenance carried out.

Simulation/emulation[16] has a number of different tasks. Relevant for AGV applications are mainly the following three simulation types:

The functional system simulation simulates the influence of changes in the AGV system. Due to a close coupling to the control functions, estimates of the system behaviour can be carried out with a high degree of implementation reliability. However, real-time simulations or simulations in individually definable time cycles are rarely available here. The basis for system simulation is the master controller itself; this procedure is also called emulation.

Material flow simulations are carried out by means of complex and abstract program packages in which the focus is on the actual logistics task. The individual transport system—in this case AGVs—is only a means to an end here. With commercial simulation systems, any type of simulation can be carried out, the software is independent of the respective means of transport and the transport task. Processes in time-lapse mode are standard here so that long-term assessments can also be carried out.

A special and so far rather rarely used type of simulation is used for project planning support. It offers functions for the rough design and determination of the functional scope of an AGV as well as for the determination of the principle system processes. This simulation component is intended to support the technical sales department during the

---

[16] See also VDI Guideline 2710-7 on the subject of "Areas of application of simulation for Automated Guided Vehicles (AGVs)".

bidding phase, but it can also already be used to generate some behavioural rules and system definitions. This information can then be used for an initial full simulation of the system.

Finally, the following should be pointed out once again: The term *AGV master control system* is often associated with a "control room", i.e. a room in which one or more persons monitor the AGV system on one or more monitors and, if necessary, even have to intervene. In large systems, i.e. spatially extensive, with a complex layout and with many vehicles, it may make sense and be helpful for the service personnel to have such a room— but it is not necessary for the operation of an AGV! Also, a modern AGV master control system does not need its own computer, but runs as software on a (virtual) server on some computer at the operating company or possibly even in an external computer centre of a service provider.

2.3 The Automated Guided Vehicle

As diverse as the applications of automated transport vehicles are, so are the designs. The range with regard to individual criteria is enormous, for example, this concerns

- the size, weight and load capacity of the vehicles,
- the number of vehicles in a facility,
- the complexity of the system in terms of functions, control, various navigation options, load handling,
- the different, sometimes extreme operating conditions,
- the different industrial sectors.

Now we do not want to capitulate in the face of this diversity, but will take a chance on a categorisation in the first section. Then we will look at some of the most important vehicle components, namely the vehicle control system, the mechanical components and the energy supply of the vehicles. We have already described the two core functionalities of navigation and safety.

2.3.1 AGV Categories

The load that is to be transported can be categorised. If we restrict ourselves to intralogistics, then the pallet (as a Euro pallet or special shapes) is certainly in the foreground. But trailers, roll containers or rolls (paper or metal) also want to be moved regularly. The following table categorises the world of AGVs (Table 2.6):

The first eight categories are unrestrictedly common in today's world of AGV. The remaining two categories are not quite so self-evident or self-explanatory.

Table 2.6 Categories of driverless transport vehicles

Cat.	Naming	Typical load	Description
1	Forklift AGV, specially designed	Pallet	Floor-level load pick-up, different heights, standard or special pallets or other forkable containers, stackable, typical load weight:1 t; designed, engineered and built by the AGV manufacturer
2	Forklift AGV as an automated series device	Pallet	As 1, but: the AGV manufacturer uses a series device from a forklift truck manufacturer and automates this with the necessary AGV technology
3	Piggyback AGV	Pallet	Usually limited to one transfer height (e.g. 1 m), lateral load pick-up by roller conveyor or chain conveyor, typical load weight: 1 t
4	Towing Tractor	Trailer	Tugger; pulls several trailers, typical total weight of trailers: 5 t
5	Underrun AGV	Roll container, trolley	The standard AGV in hospital logistics, among others: it drives under the roll container and lifts it for transport, typical load weight: 500 kg, vehicle height approx. 350 mm More recently, increasingly also in industrial applications with up to 1.000 kg or even more load weight and (max.) 220 mm vehicle height
6	Assembly AGV	assembly object	Use in series assembly: the mounting for the assembly object sits on a substructure, typical load weight: up to 1 t
7	Heavy duty AGV	rolls, coils (paper or metal)	Transport of heavy paper rolls or steel coils up to 35 t; transport of (deep sea) containers, workpieces/components etc.
8	Mini-AGV	bins/boxes	Use in larger fleets, e.g. for order picking
9	Outdoor AGV	various	Usually outdoor vehicles, mostly diesel-electric or diesel-hydraulic drive, typical load weights $\geq$ 3 t. Examples: Diesel forklift, truck, wheel loader, harbour AGV for deep sea containers
10	Special Design AGV	various	Special solutions for special tasks; all AGVs that do not fit into one of the above ten categories

KLT—German acronym for small load carrier, various containers (boxes, bins, totes) for small parts

We would like to point out at this point that we are excluding the entire topic of "service robotics" here. Service robotics will become increasingly important—that much is certain, but

- so far—we are looking at the third AGV epoch here—the importance of service robotics for the AGV world has been very low and continues to lag well behind the expectations of the 1980s and 1990s, and
- here we describe the use of the AGV in intralogistics, where the vast majority of service robotics applications do not belong to.

2.3.1.1 Forklift AGV: Specially Designed

The range of applications for these vehicles is wide. The focus is on the pallet or the forkable container (Fig. 2.27). The logistical tasks can be very simple (simple transports between two locations without many branches), but also complex (taxi operation). The vehicles can be used as independent vehicles (stand-alone) or, managed by an AGV master control system, they can work in a network with (many) other AGVs.

2.3.1.2 Forklift AGV as an Automated Series Equipment

In principle, the operational spectrum of these vehicles is similar to that of Cat. 1.

Essential is the use of series vehicles from the standard series of forklift truck manufacturers, which are automated with as little intervention as possible. The AGVs shown in Fig. 2.28 can still be operated manually after automation—therein lies its particular advantage. There are pedestrian and ride-on vehicles.

The series forklift truck is supplemented by the safety devices, the control system and the navigation components. For dead reckoning, for example, the drives must be equipped with incremental and rotary encoders. In modern series devices, electronic modules and bus systems are used as standard to control the drives, so that they are already well prepared for automation.

There are two different camps in the AGV world: some prefer specially designed vehicles, others see the advantages in automated series equipment. The advantages of specially designed AGVs (Cat. 1) are:

- Optimal integration of the additionally required components (space problem),
- Designed for continuous use and extended service life,
- Consideration of an energy concept suitable for automation (automatic battery change or charging).

The advantages of the automated series units are cited as follows:

- Cost advantages due to series production,
- Proven service and cost-effective spare parts stocking.

When automating series devices, serious AGV manufacturers will always check whether components do not need to be replaced for years of continuous use in the AGV (this includes wheels and electrical components, for example).

Fig. 2.27 AGV cat. 1: Forklift AGV—specially designed. (Sources: left DS AUTOMOTION, right Rocla)

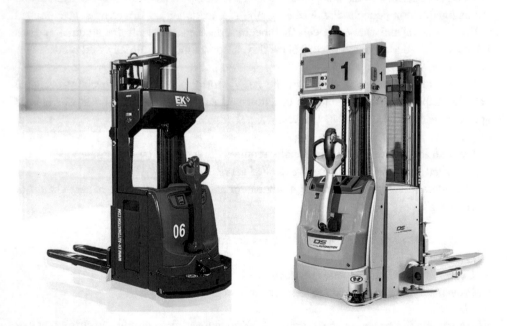

Fig. 2.28 AGV cat. 2: Forklift AGV, basis: series truck. (Sources: left E&K Automation—Linde L14, right *DS* AUTOMOTION—Still EXV14)

Fig. 2.29 AGV cat. 3: Piggyback AGV. (Source: Oceaneering)

2.3.1.3 Piggyback AGV

These vehicles also transport the classic loading aids such as the pallet, the containers or skeleton containers. In contrast to the two aforementioned categories, however, piggyback AGVs cannot pick up the loading equipment from the ground, but require a certain height of usually more than 60 cm, which must then be maintained throughout the entire facility as the standard transfer height—leaving aside elaborate mobile or stationary equipment to adjust the transfer height (Fig. 2.29).

The great advantage of these vehicles lies in the load handling: the lateral load pick-up makes it possible to drive directly to the stationary load transfer point without manoeuvring—as the fork lift vehicles have to do—and to take over the load by means of the conveyor (roller conveyor, chain conveyor or similar). This is done quickly and requires less space.

2.3.1.4 Towing Tractors

Strictly speaking, the specially designed AGVs and the automated series trucks must be distinguished here, too. However, since the significance of trailer tractors in terms of unit numbers is much lower than that of forklift AGVs, we have dispensed with separate categories (Fig. 2.30).

Fig. 2.30 AGV cat. 4: 2 towing tractor. (Sources: left E&K Automation, right dpm)

Here, too, with the automated series trucks, the facilities for manual operation (driver's platform with steering wheel and accelerator pedal or joystick) are usually retained and thus—temporary—manual operation remains possible without any problems.

The safety aspect is less problematic with this type of AGV: when a tugger train consisting of a towing vehicle and three to four trailers sets off, there is a danger zone in all gaps that can hardly be monitored by sensors. Therefore, before the start of an automatic drive—and this also means after a stop due to an obstacle in the drive track—it is usually necessary for an operator to confirm by pressing a button that all gaps are free of people and that driving off is possible without endangering people.

2.3.1.5 Underrun AGV

In the case of underrun AGVs, a distinction can essentially be made between the two fields of application "hospital" and "industry". Due to the hygiene standards that have to be met in hospital applications, the chassis of these AGVs are made entirely of stainless steel—which is rather unusual for an industrial AGV and may only be required for use in the food or pharmaceutical industry.

Both applications have in common that the vehicles drive completely under the provided load and then lift it a few centimetres by means of an integrated lifting mechanism so that the wheels of the transport containers are free of the ground. In this way, a significantly better track stability is achieved, as the condition of the wheels—damaged surface of the wheel bandages, stiff wheel and steering bearings—has no influence on the driving behaviour of the AGV (Fig. 2.31).

The load can be picked up automatically—supported by corresponding sensors on the vehicle—or manually by pushing the roll container over the vehicle ready for collection. Setting down the transport container at the destination is almost always done automatically.

Another application for underrun AGVs that has become established in recent years is their use in distribution centres to support the "goods-to-person" picking principle. In this case, the goods are arranged in approx. 2 m high shelving racks, which have a square or rectangular base area of about 80–100 × 80–100 cm and can be driven under and lifted by

Fig. 2.31 AGV cat. 5: Underrun AGV. (Sources: top left Swisslog, top right and bottom left DS AUTOMOTION, bottom right E&K Automation)

the AGV (Fig. 2.32). The racks stand close together on a staging area and are brought by the AGVs through narrow aisles into the picking area. After picking the item(s), the rack is returned (by the same vehicle) to the staging area, but not necessarily to the same place as before, as the entire warehouse is constantly optimised with regard to the picking frequency of the stocked items: each time a rack is returned, the warehouse management system decides anew where (= at what distance from the picking zone) it should be placed.

2.3.1.6 Assembly AGV

AGVs used in assembly lines differ considerably. Here, the assembly object with its size and weight essentially determines the vehicle. But the assembly steps also play a role in the design of the AGV: Are purely manual assembly steps planned or are there also automatic stations? This results in different requirements for positioning accuracy, for example. And: How great are the forces that act on the object and thus on the AGV during the assembly activity? This results in different requirements for tilt stability. In addition, the required accessibility to the assembly object must be ensured.

In most cases, such systems are simpler in terms of control than taxi systems. The driving speed is extremely low and the requirements for workers' safety and protection are often different. Workers are constantly in the immediate vicinity of the vehicles. They should be able to go about their work without restrictions, but still be protected from injury. This often has an impact on the use of a safety laser scanner, on the setting of its protective and warning fields and on the lateral step protection. The safety design must be in such a

Fig. 2.32 AGV cat. 5: 3 Underrun AGV. (Sources: top left Swisslog, top right Grenzebach, bottom Amazon Robotics)

way that the worker can work safely, but still the sensor system does not continuously respond without reason (Fig. 2.33).

2.3.1.7 Heavy Duty AGV

Here we want to group those heavy-duty vehicles that operate indoors. Examples are vehicles that transport rolls, either in the paper producing or processing industry (paper rolls weighing several tonnes) or in the steel industry as producers of steel coils or the automotive industry as their consumers (steel coils usually weigh up to 30 t) (Fig. 2.34). However, large/heavy machine parts can also be transported or high/long/heavy containers for large quantities of solid or liquid substances (also Fig. 2.34).

Vehicles for such load weights place high demands on all designs and components. This applies to the drives, the energy supply and the safety technology. It is in the nature of things that the number of applications for such extreme weights is comparatively low.

The high weight of the entire transport determines the efforts of the developers to avoid accidents with people or things under all circumstances. This sounds obvious at first, so we should explicitly point out the direct comparison to AGV category 6 here, which is also particularly concerned with safety design—but with a completely different objective.

Fig. 2.33 AGV in assembly; top left at Daimler in Bremen (Source: CREFORM), top right at Rolls Royce Power Systems in Friedrichshafen (Source: dpm), bottom left at BMW in Berlin (Source: DS AUTOMOTION), bottom right at Fendt in Asbach-Bäumenheim (Source: DS AUTOMOTION)

2.3.1.8 Mini-AGV

The eighth category is still rather rare in the third AGV epoch. Here it is a matter of many small and inexpensive vehicles carrying out a large number of transports—usually bins or boxes/cartons of comparable size with weights of up to 25 kg (Fig. 2.35).

The demand for a low unit price is solved with the help of intelligent designs and new components (electrics, electronics and sensor technology). A sticking point here is again the protection of persons, because today's certified safety laser scanners approved for the protection of persons would go beyond any set price frame. One solution for the use of inexpensive sensor technology is described by the following argumentation, which is also accepted in this way by the BG[17]: if the vehicles (incl. load) are not too heavy and also do not move particularly fast, the possible injury in the event of a collision with a person is rather low and in no case fatal. For this reason, sensors whose so-called performance level (PL) is below the otherwise required level "d" may also be used with these AGVs. For example, a safety laser scanner with performance level "b" also stops the AGV in time before a collision with a human, but it has no built-in self-monitoring of its proper function

---

[17]BG = German acronym for *German Social Accident Insurance Institution.*

Fig. 2.34 AGV Cat. 7: Heavy duty AGV. (Sources: top left Frog/Siemag; Frog today belongs to Oceaneering, Siemag to AMOVA; top right MLR, bottom Alcoa Mosjøn)

Fig. 2.35 AGV cat. 8: Mini-AGV. (Sources, from left to right: BITO—"Leo Locative", Götting— "KATE", SSI Schäfer—"Weasel")

and its theoretical probability of failure is somewhat higher than that of a more expensive device with PL-d. Therefore, it is theoretically possible that the device fails unnoticed and as a result the AGV does not stop in time before an obstacle, but collides with it.

Many unconventional applications are conceivable for these Mini-AGVs. The best-known task is in an advanced type of order picking where the employee does not search for and collect the goods, but the goods come to the employee independently—with the help of the Mini-AGV—and give him assistance in putting together the customer-specific deliveries. But also the supply and disposal of production workplaces, e.g. in electronics manufacturing, can be realised inexpensively with these Mini-AGVs.

It is a vision that no longer only researchers in universities are working on: Entire "swarms" of small AGVs work intelligently with each other—similar to how ants or bees, for example, carry out transports collectively. The vehicles will communicate with each other, develop strategies and solve tasks together—even without their own separate AGV control system. The associated research areas are called multi-agent systems and swarm intelligence, but require a lot of sensors or sensor data on the vehicle side as well as considerable computing power for their evaluation—this then no longer corresponds to the vehicle characteristic "inexpensive" mentioned at the beginning. At this point, we refer to Chap. 4 and the section on autonomous AGVs.

2.3.1.9 Outdoor AGV

In this category we summarise different vehicles that are used outdoors. These are mostly larger AGVs that transport loads of several tonnes. Because of the large vehicle and load weights, combustion engines are often used, which then enable diesel-electric or diesel-hydraulic propulsion,However, purely electric outdoor AGVs are also in use (Figs. 2.36 and 2.37).

The special features of outdoor use are described in detail in Sect. 3.3 "Areas of application—outdoor use".

2.3.1.10 Special Design AGV

In this category we collect all vehicles that have been specially designed and built for very specific projects, i.e. that do not fit into one of the front categories (Fig. 2.38).

2.3.2 Vehicle Control

In the classic hierarchical control structure of the third AGV epoch, the vehicle control is subordinate to the AGV master control system. While the master control system is concerned with the whole, namely the fulfilment of the task (efficient processing of transport orders), the vehicle control coordinates all actions within the AGV.

The vehicle control, consisting of hardware and software, is thus one of the most important components in an AGV. The hardware can be very different, depending on the complexity of the system and the intelligence of the vehicles:

- Single-board computer,
- programmable logic controllers (PLC),
- custom-made integrated computers based on micro controllers (Fig. 2.39),
- multi-board computer,
- standard industrial PC (IPC) with fieldbus interfaces.

Fig. 2.36 AGV cat. 9: 2 Outdoor AGV, left: Wheel Loader right: in a container port. (Source: Götting)

Fig. 2.37 Top left: Automated truck with diesel drive, tactile bumper and laser scanner (Source: Götting), top right: follow-up project at the same user, AGV with electric drive and without driver's cab (Source: Kamag); bottom: Underrun AGV for tank transport (Source: BASF SE)

2.3.2.1 Requirements for a Vehicle Control System

The use of the control unit in a vehicle results in a number of special requirements. For example, it must be able to cope with voltage fluctuations of the mobile power supply, be protected against penetrating moisture and dust particles, and function reliably even with

Fig. 2.38 AGV cat. 11: Special Design AGV. (Source: Hencon; world's first AGV for use in the electrolysis hall (potroom) in aluminium smelting electrolysis with extremely strong electromagnetic fields)

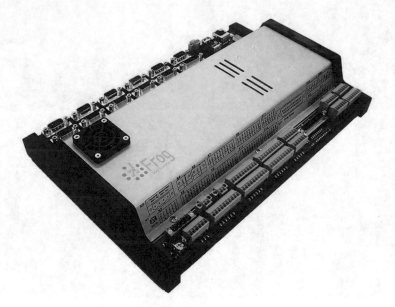

Fig. 2.39 A custom-made AGV control unit. (Source: Frog/Syslogic)

strong vibrations and shocks. If the vehicle is to be used outdoors, there may also be increased requirements due to the climate. Under such conditions, however, the control units are usually installed in a shielded location and cooled and/or heated.

The vehicle control system is of central importance when it comes to safety requirements. In accordance with the required high safety categories according to EN 954 or the safety integrity levels of IEC 61508 for the safety and protection of persons, the functional integration of the vehicle's electrics/electronics must be carried out in a complex manner:

- Hardware and software of the AGV control system (personal safety must be guaranteed),
- selection of sensors plus evaluation unit ($\rightarrow$ fail-safe),
- design of the electrical system ($\rightarrow$ 2-channel).

2.3.2.2 Vehicle Control Interfaces

Today, the interface between the vehicle control and the master control is technically mostly a WLAN data transmission and logically the driving order, which comes into the vehicle via this physical interface and is reported back in the same way after completion. The interfaces of the vehicle control within the AGV concern the following vehicle components:

- Safety system, i.e. emergency stop circuit, safety laser scanner for persons' safety and protection, safety bumpers etc.,
- energy management, i.e. monitoring the battery charge levels,
- load handling device, i.e. the position of a lifting fork or similar,
- mechanical drive elements for driving and steering,
- operating devices, i.e. the control panel (Fig. 2.40) and the hand control unit.

There may also be interfaces to external devices or equipment:

- Direct communication to other vehicles,
- load lifting stations,
- building equipment such as lifts, lifts, automatic doors, fire section gates, traffic lights, barriers, etc.

2.3.2.3 Typical Function Blocks

The headline already limits the function blocks considered here, because the functional distribution of tasks between the master controller and the vehicle controller can be realised very flexibly. In extreme cases, it is conceivable that all function blocks of both control hierarchies run in the vehicle computer. However, this is still rather uncommon today; therefore, we will limit ourselves here to the classic division of the function blocks, as shown in Fig. 2.41.

The **Manager** function block breaks down the driving task into individual commands and uses the vehicle components such as the drives, steering, load handling devices, safety devices, etc. to fulfil the driving task. It is therefore equivalent to a control centre in the AGV.

The **driving** function block applies the navigation procedures by combining dead reckoning and bearing with the mechanical driving devices such as drive and steering

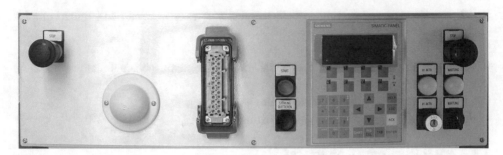

Fig. 2.40 A control panel on an AGV with: 2 black stop buttons (*no* emergency stop!), a round WiFi antenna, the connection socket for the hand control, an input terminal, buttons and lights. (Source: dpm)

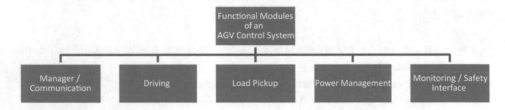

Fig. 2.41 The function blocks of a typical AGV control system

actuators in accordance with the options described above. All tasks of position determination, navigation and track guidance are carried out in this block.

The **load pick-up** function block takes over the coordination of load pick-up and delivery. Depending on the complexity of the LHD,[18] the sensors are queried and the actuators are controlled. If the stationary load transfer device is also active, this function block also takes over its control.

The **energy management** function block has the task of keeping the vehicle's energy system operational with maximum availability. Now, the possibilities of mobile energy supply are manifold; in principle, all combinations of the following technologies are conceivable:

- accumulators, such as lead-acid/lead-gel and NiCd,[19]
- alternative accumulators, such as nickel-metal hydride (NiMh), lithium-ion (LiIon),
- contactless energy transmission via induction,
- double layer capacitors,
- fuel cell,
- combustion engine drives, such as petrol or diesel.

---

[18]LHM = Load Handling Device.
[19]NiCd = Nickel Cadmium.

Depending on the technology used—combinations are also conceivable—energy management can be implemented with very different levels of complexity.

The **monitoring & safety** function block guarantees the protection of persons and property, which is why its functional fulfilment is at top priority. Safe controls and/or a safe circuit design are required to guarantee the function of the personnel protection devices in any case. Lower safety requirements are placed on the second functionality of this function block: collision prevention. It is part of the traffic control and ensures that several AGVs of an AGV system neither collide nor block each other. For this purpose, the vehicles are either equipped with appropriate collision prevention sensors or they communicate directly with each other.

2.3.2.4 Operating Modes

The control structure of a vehicle control can be different depending on the selected/activated operating mode. In addition to the operating mode "automatic mode", there are other operating modes such as

- manual operation,
- semi-automatic operation,
- diagnosis & Service,
- learning mode.

Manual operation enables the operation of the vehicle functions via the control panel on the vehicle or with the manual control unit. The manual control unit is an external device that can be connected to the AGV by cable and plug or by radio communication, with which the AGV can be moved, for example, by joystick. In some cases, it is also possible to pick up or set down the load manually with this device.

Semi-automatic operation is always a project-related specific solution. It is usually a mixture of automatic and manual operation. For example, if the AGV master control system fails or the WLAN connection between the vehicles and the master control fails, it may make sense for the vehicles to continue driving (with restrictions/limitations if necessary). For this purpose, the driving orders could be entered manually on the control panel and automatically executed. However, if interfaces to the environment of the AGV are not operated directly by the vehicle, but are handled via the AGV master control system, this intention quickly reaches the limits of its feasibility.

The operating mode Diagnosis & Service allows far-reaching interventions and operating options for the service personnel.

In the learning mode, new drive track information is taught in. This can be done directly by means of one-off manually executed drives (teach-in) or by downloading from the AGV master control system. These journeys are stored in the vehicle computer and are then available in automatic mode.

2.3.3 Mechanical Movement Components

As diverse as the vehicle categories described in Sect. 2.3.1 are the technical solutions used to enable vehicle movement. This requires the wheels (of course, we only consider "wheeled vehicles"), the chassis—i.e. the number, type and arrangement of the wheels—as well as the drives and steering.

2.3.3.1 Wheels

Most AGVs—especially almost all indoor vehicles—have wheels with bandages made of plastic (elastomers), mostly Vulkollan® (Bayer) or polyamide. They have a high abrasion resistance and leave little abrasion on the road surface (not "chalking").

For outdoor vehicles, you will also find solid rubber tyres or ordinary truck tyres (pneumatic rubber tyres). The more elasticity the tyres get, e.g. in order to score in terms of comfort (possibly important for the on-board electronics and/or the load) and/or off-road capability (in poor road conditions), the more demanding the path control of the vehicle becomes and the worse the positioning accuracy.

2.3.3.2 Wheel Configuration

The wheel configuration describes the selection, number, arrangement and control of the wheels of an AGV. If you increase the effort in the wheel configuration, you gain in the mobility of the vehicle. Good manoeuvrability means less space required and time saved when driving on a straight line, in curves, when manoeuvring and when changing loads. However, anything that brings advantages for the application causes additional costs.

The optimal chassis is derived from the technical necessity and the economic efficiency of the project. The space requirement can be assessed with the help of envelope analysis, which graphically displays the total area covered by the vehicle. In this way, the specified layout can be checked in advance with regard to the spatial situation at the load pick-up and drop-off points as well as at curves and narrow points in the drive track. If it is possible to make the vehicles faster and thus more efficient with the help of a technically sophisticated running gear, the higher vehicle price can be offset by reducing the required number of AGVs.

Figure 2.42 shows typical wheel configurations. A distinction is made between line-moving vehicles with two degrees of freedom of movement and omni-directional vehicles with three degrees of freedom of movement. The envelope curve is less favourable (larger) for line-moving vehicles than for omni-directional ones. Thus, the three-wheeler chassis is comparable to the familiar movement behaviour of a car, of which everyone knows that there is overshooting when cornering or parking and must be taken into account. If our car had a omni-directional chassis with all-wheel steering, parking would be easier for everyone.

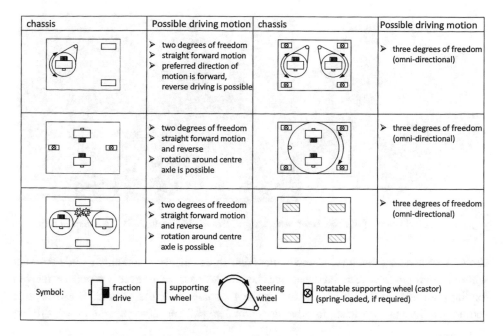

chassis	Possible driving motion	chassis	Possible driving motion
	➤ two degrees of freedom ➤ straight forward motion ➤ preferred direction of motion is forward, reverse driving is possible		➤ three degrees of freedom (omni-directional)
	➤ two degrees of freedom ➤ straight forward motion and reverse ➤ rotation around centre axle is possible		➤ three degrees of freedom (omni-directional)
	➤ two degrees of freedom ➤ straight forward motion and reverse ➤ rotation around centre axle is possible		➤ three degrees of freedom (omni-directional)

Symbol: fraction drive   supporting wheel   steering wheel   Rotatable supporting wheel (castor) (spring-loaded, if required)

Fig. 2.42 Sketches of typical chassis of AGVs. (Source: VDI 2510)

A special chassis becomes possible with the Mecanum wheel.[20] This enables a vehicle with amazing maneuverability that can perform any conceivable movement in the plane from a standing position. The three or four wheels of a vehicle operate without geometric steering lock. Above the circumference of the wheel are several individually pivoted "barrels" that can rotate freely. Each Mecanum wheel has its own drive, the speed of which must be regulated exactly according to the specifications of the vehicle control system. The superimposition of the three or four—identical or also different—rotational speeds results in the respective resulting movements of the vehicle (Fig. 2.43).

Another classification criterion is the number and arrangement of the wheels:

– Three-wheeler chassis	→	Triangle shape
– Four-wheeler chassis	→	Rectangle or rhombus
– Five-wheeler chassis	→	Gable form
– Six-wheeler chassis	→	Rectangle shape.

To increase the load-bearing capacity of a vehicle, it is always possible to attach additional support wheels in addition to the required functional wheels, which usually with suspension—help to lift the total weight consisting of vehicle and load weight. It

---

[20]The Mecanum wheel was invented in 1973 by Bengt Ilon, an employee of the Swedish company Mecanum AB.

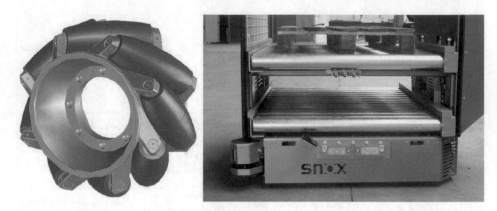

Fig. 2.43 Left: Sketch of the Mecanum wheel, right: Installation in an AGV. (Source: Snox)

should be noted here that a vehicle with more than three wheels is statically overdetermined without appropriate additional measures, i.e. in the case of uneven ground or uneven loading, not all wheels necessarily always have contact with the ground. However, this should be the case at least for the driven/braked wheels always and in all driving conditions!

An additional wheel can also have the function of a measuring wheel. Such a smooth-running wheel is spring-mounted and does not transmit any weight forces or drive or steering forces. Its sole purpose is to measure the movements of the vehicle in order to provide the dead reckoning algorithm with values that are as unadulterated as possible.

2.3.3.3 Steering
Here we only want to point out a difference in principle: There are steering systems with and without geometric steering angle.

An example of the geometric steering angle is a three-wheeler chassis with a steered and driven front wheel and two trailing fixed wheels on the rear axle.

Running gears without geometric steering angle are, for example, the differential or speed differential drive as well as a running gear based on three or four Mecanum wheels.

2.3.3.4 Drives
The voltage range of the electric drives extends from 24 to 96 V. Both DC motors as well as maintenance-free three-phase drives are used. Especially wheel hub drives with so-called brushless DC motors (BLDC motor = Brushless DC Motor—contrary to the designation, actually a three-phase motor with excitation by permanent magnets) are becoming increasingly popular, as they are more expensive to purchase but maintenance-free and therefore cheaper to operate.

Of course, all AGVs are equipped with mechanical, hydraulic and/or electric holding brakes.

A typical traction drive consists of an electric motor, a wheel hub gear, a running wheel with Vulkollan® bandage and an electromagnetic brake (Fig. 2.44).

Fig. 2.44 Typical wheel hub drive for AGVs: The RNA 27 with integrated steering unit, 270 mm wheel diameter, 1300 kg wheel load, 24 or 48 V, available in DC or AC technology. (Source: Schabmüller)

We will refrain from a detailed consideration of internal combustion engine drives for outdoor vehicles here. In most cases, these are diesel-hydraulic or diesel-electric drive concepts, i.e. the combustion engine only drives the AGV indirectly by supplying a hydraulic system with pressure or by driving an electric generator, which in turn supplies electric drives with energy.

2.3.4   Energy Supply of AGVs

Automated guided vehicles must be supplied with energy for

- the vehicle control system, the electrics, electronics and sensors,
- the traction and steering drives and
- the devices for load handling.

Outdoor vehicles usually have—similar to trucks—both a battery (usually lead-acid accumulator) for the electrical components and a tank for gas, petrol or diesel. We will not go into this in detail here. Under certain circumstances, indoor AGVs can be supplied with electrical energy using conductor lines, but this is unusual and will therefore not be pursued further.

At this point, we want to focus on the three common technologies used to supply energy to AGVs:

1. traction battery (lead-acid, NiCd, lithium-ion accumulators)[21]

---

[21] Lithium-ion batteries have had the highest growth rates in the global battery market for years.

2. contactless energy transmission
3. hybrid system: contactless energy transfer plus (small) back-up battery or double-layer capacitor (so called PowerCap)

All these techniques have their justification; we want to look at them individually and then also distinguish them from each other or limit their suitability. Table 2.7 compares the possibilities so far.

Note: "Lithium-ion" is used as a collective/generic term. They come in many varieties that differ in the electrode material used (anode, cathode). Depending on the combination of electrode materials, batteries result that are better or worse suited for different applications.

Examples of cathode materials: lithium cobalt oxide (LCO or LiCoO2), lithium manganese oxide spinel (LMS, LMO or LiMn2O4), lithium nickel cobalt manganese (NMC, NCM or LiNiCoMnO2), lithium iron phosphate (LFP or LiFePO4), lithium nickel cobalt aluminium oxide (NCA or LiNiCoAlO2); anode materials: graphite, lithium titanate oxide (LTO or Li4Ti5O12), silicon and silicon-carbon composites, pure lithium (lithium metal anode and polymer electrolyte).

2.3.4.1 Traction Battery
Traction batteries commonly used in AGVs today:

- Lead-acid batteries (liquid electrolyte),
- lead-gel or lead-fleece batteries (bonded electrolyte),
- lithium-ion batteries.

The choice of battery depends, among other things, on the operating mode of the vehicles. The following battery operating modes are common for AGVs:

(a) Capacitive discharge with and without battery change
(b) Capacitive discharge with intermediate charges
(c) Cycle operation ("Opportunity Charging")[22]

(a) Capacitive discharge (lead battery)
   Capacitive discharge requires that the battery is fully charged at the beginning of the work shift. The battery capacity is dimensioned in such a way that the operating capacity, i.e. max. 80% of the nominal capacity, is sufficient for the planned operating time (usually one work shift). During the discharge process, the limit values must not be

---

[22]Opportunity charging: every opportunity that arises during the process is used for intermediate charging.

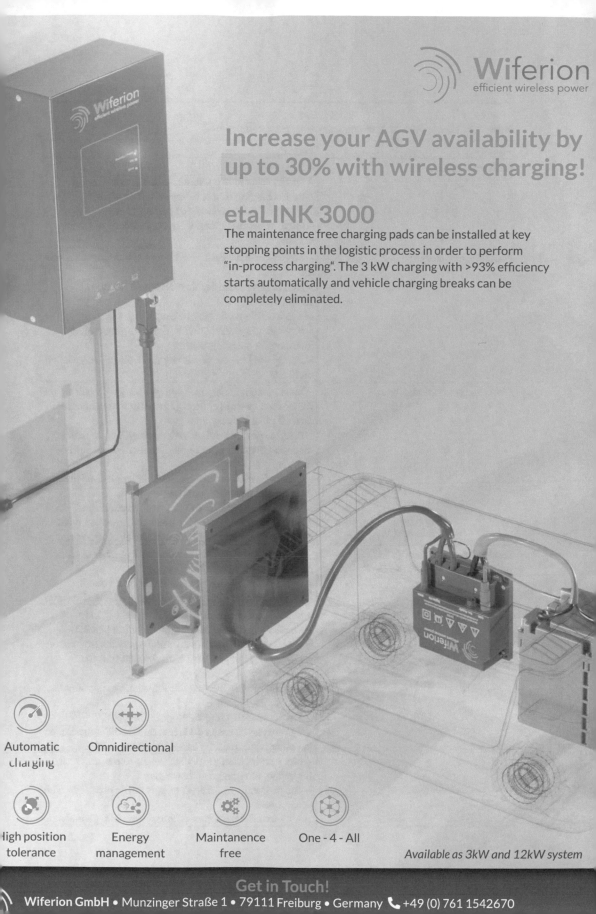

Table 2.7 Comparison of common energy supply technologies

Technology	Property
Lead accumulator	– Inexpensive – Use in capacitive operation (see below) – Long charging times, therefore either AGV "overnight" to the charger or battery change – High weight as counterweight or advantageous for stability
NiCd accumulator	– More expensive than lead batteries, but longer service life – Higher power density than lead-acid batteries, i.e. smaller – Use in cycle mode (see below) – Fast-charging capable with high charging currents; intermediate charging possible – Hardly used in new plants any more due to heavy metal problems
Lithium-ion accumulator	– Most modern battery technology currently available on the market; currently still significantly more expensive than all other battery types – Positive properties like NiCd battery, but no heavy metal; no memory effect with frequent intermediate charging – Cannot be used at very low and very high temperatures (outdoor) or air-conditioning is required. – Development continues with new electrode and electrolyte materials with the goals of higher cell voltage, higher capacity and a greater number of charging cycles
Double-layer capacitor (DLC) also: "PowerCap", "SuperCap"	– Wear-free, i.e. almost infinitely high number of charge/ discharge cycles – Energy density significantly lower than batteries – Not suitable as sole energy storage ($\rightarrow$ hybrid system)
Contactless energy transmission	– Suitable for simple layouts, e.g. assembly lines – Smaller space requirement than traction batteries – Wear- and maintenance-free and operationally reliable – Dispensing with traction batteries and thus their disadvantages: – Limited service life, regular replacement investments, effort for maintenance – Environmental protection: gases, disposal – System can be easily switched on and off, batteries, on the other hand, must be cared for and maintained – Can also be used locally/stationary to transmit energy for battery charging, is advantageous: – The positioning of the vehicles is not as precise as with contact-based energy transfer necessary (similar to mobile phone

(continued)

Table 2.7 (continued)

Technology	Property
	charging cradles in the consumer sector) – Approach from any direction or with any orientation/ rotation possible – Waterproof encapsulation of the external components (IP67), no abrasion ($\rightarrow$ suitable for applications in clean rooms, in the pharma-ceutical and food industries)
Hybrid system: contactless energy transmission + Xxx accumulator	– Back-up battery or double-layer capacitor helps to cover short-term power demand – Back-up battery or DLC prevents failure if the position of the vehicle above the double electric conductor is not optimal, e.g. in curves or when manoeuvring
or Contactless energy transmission + DLC	– Backup battery or DLC as energy source in case of power failure or in manual mode off the lane
Fuel cell (hydrogen-based)	– Currently still very rare – Expensive infrastructure: Hydrogen tank in the AGV and hydrogen filling station (indoor) for the AGV + hydrogen tank (outdoor) for gas storage. – Backup battery required to buffer power peaks (e.g. high currents when starting or lifting heavy loads) – Lifetime currently still too short (especially for 24/7 use)

exceeded (current, temperatures, etc.). After the operating capacity has been extracted, there must be enough time to fully charge the battery. As a rule, the same time is needed for charging as for discharging or driving (at least 7.5 h). Recharging takes place either directly in the vehicle (mostly single-shift operation) or—including the manual or automatic battery change—in an extra battery charger room (mostly multi-shift operation).

These systems are the simplest and cheapest of those considered here. They can be used from 1-shift to 3-shift operation, but they require a relatively large amount of effort for changing and maintenance in multi-shift operation. Furthermore, in case of battery change and charging outside the vehicle, (at least) one additional battery must be provided for each AGV.

(b) Capacitive operation with intermediate charging (lead battery)

Capacitive discharge with intermediate charges also assumes that the battery is fully charged at the beginning of the work shift. The battery capacity is dimensioned so that the operating capacity, i.e. max. 80% of the nominal capacity including the sum of the capacity increases due to recharges, is sufficient over the entire work shift. Recharging during breaks in operation lasting 30–60 min increases the energy turnover of the batteries and causes higher temperatures in the battery, which reduces the service life of the battery. Here, too, recharging usually takes at least 7.5 h. This operating mode is usually only found in 1-shift operation, rarely in 2-shift operation.

(c) Cycle operation (Li-Ion and, with some restrictions, also lead-acid batteries)

For cycle operation, the operating capacity of a battery is designed so that the energy reserves are sufficient until the next charging time. The charging of the battery takes place in or next to the plant, usually even during a work process (opportunity charging). At the charging location, there must be enough time to recharge the energy that has been taken out. The batteries remain in the vehicle for the entire working day and are only charged by intermediate charges. There is no need to recharge or change the battery every day. The daily energy throughput is decisive for the dimensioning. The lithium-ion battery variants that are particularly suitable for cycle operation are maintenance-free and are usually used in 3-shift operation. With this system, round-the-clock operation (24 h a day and 7 days a week) is common or possible. Although the initial investment is higher than with the previously mentioned systems, the total costs (consideration of the utilisation time of the AGV, also called TCO—"total cost of ownership") are lower due to a longer service life and lower maintenance requirements.

2.3.4.2 Contactless Energy Transmission

In contactless energy transmission, electrical energy is transmitted inductively from a conductor laid firmly in the ground to one or more mobile consumers (AGV) without contact. The electromagnetic coupling takes place via an air gap and is maintenance- and wear-free. The primary circuit consists of only one winding, which is permanently installed in the ground as a "double conductor" along the drive track of the AGV. About 2 cm above the ground, the secondary circuit then sits in the AGV, where the induced energy is made available to the consumers in the vehicle. The transmission frequency is usually 20–25 kHz.

This method of supplying the AGVs with power is suitable for simple AGV layouts, such as those found in series car assembly. For complex layouts where the AGV is in taxi operation, this is difficult to realise.

We will not go into the technical basics[23] here. Instead, it is worth taking a look at the necessary components and the installation of the double electric conductor.

The mobile components of such systems include the transformer head (secondary part, also called pickup) and the matching actuator connected to it. Several transformer heads can also be installed on the underbody of the AGV in order to be able to transmit the required power. A common value for the power of one head is 800 W. The matching

---

[23] Further reading book by Dirk Schedler: "Kontaktlose Energieübertragung—Neue Technologie für mobile Systeme", Verlag "Die Bibliothek der Technik", ISBN 978-3-937889-59-7.

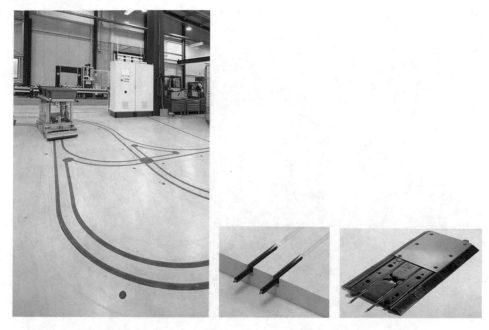

Fig. 2.45 AGV with contactless energy transmission; the pictures in the middle and on the right show laying options of the double conductor. (Source: SEW)

converter then converts the induced current into DC voltages, usually a control voltage of 24 V and a power voltage of 500 V DC (Figs. 2.45 and 2.46).

The most important stationary components are the feed-in converter, the connection module and the compensation capacitors. The feed-in converter converts the input AC voltage (50/60 Hz) into an AC voltage with the specified transmission frequency of 20–25 kHz. The power is typically 16 kW. The connection module turns it into a constant sinusoidal alternating current.

In addition to this—mostly continuous—transmission by means of double conductors, there is also the possibility of placing a transformer coil on the ground at individual selected points or of embedding it in the ground, which then also enables contactless energy transmission into the vehicle in order to charge a battery. The required stationary and mobile system components are—except for the conductor pair or the coil—very similar to the previously described two-conductor technology, however, significantly higher frequencies are used here, which contributes to a better efficiency when bridging the air gap. A major advantage of this technology which is aimed in particular at applications in opportunity charging, is that the positioning accuracy to be maintained by the vehicle is rather low compared to contact-based energy transmission: The coils in the vehicle and in the ground may have an axial offset of up to 10 cm, i.e. in this way, vehicles can also be charged automatically that cannot hit charging contacts precisely enough due to their navigation and line guidance technology and the limited accuracy that can thus be

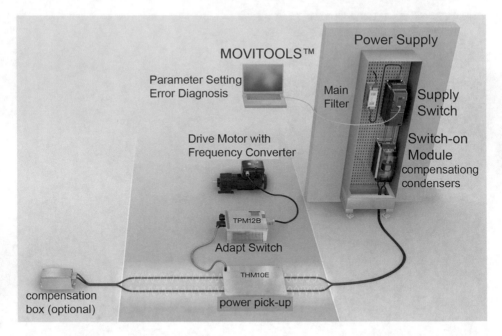

Fig. 2.46 Components of a system for continuous contactless energy transmission. (Source: SEW)

achieved. Also, unlike charging contacts, approaching the charging position is possible from any direction and with any direction/rotation (Fig. 2.47).

2.3.4.3 Hybrid System

By hybrid system we mean an energy supply system for AGVs that consists of a combination of contactless energy transmission and a so-called back-up battery. Backup battery because it can be much smaller than the typical traction battery and because it only fulfils limited tasks. There are numerous reasons why a back-up battery is useful or necessary, the main ones being:

- Not all routes of a layout can be equipped with double electric conductors. Where special requirements are made on the floor or layout flexibility, it is necessary to drive freely, i.e. without the double conductor.
- The process may require power peaks in the AGV, which are served with an additional energy storage. High-performance capacitors are of course also conceivable here.
- In the event of a malfunction, the vehicle can be removed from the layout using the back-up battery so as not to disturb the other AGVs.
- In special layout areas, e.g. in tight curves, the relative position of the primary (stationary) and secondary (mobile) circuits can be so unfavourable that the transmitted power would not be sufficient for driving alone—the back-up battery then helps.

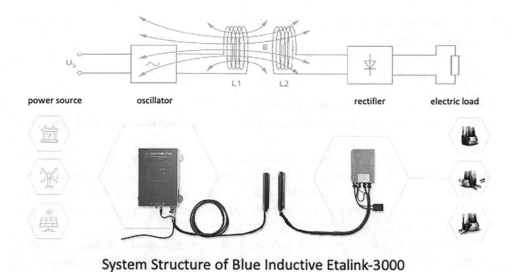

System Structure of Blue Inductive Etalink-3000

Fig. 2.47 System for local contactless energy transfer. (Source: Blue Inductive/Wiferion)

2.4 Environment of the AGV System

An AGV is **the** flexible automatic transportation technology. It can be integrated into almost any given industrial environment. This means that during the planning of the system, the stationary environment must be analysed so that the AGV can be adapted to its operational environment and appropriate interfaces to neighbouring trades can be created.

However, it must be clear to everyone that this integration has an importance that should not be underestimated. In many projects, what was neglected to be clarified in advance is repaired far too late.

2.4.1 Operating Environment

The operational environment is determined on the one hand by the environmental conditions (atmosphere) and on the other hand by spatial constraints. The atmosphere must always be taken into account as soon as it deviates from the norm; this is the case when the following conditions are to be expected:

- Particularly high or low temperatures, i.e. below 5 °C and above 30 °C,
- large temperature fluctuations,
- increased humidity or extremely dry air,

- additives in the atmosphere, such as oil mist, solvents, water vapour, paint particles, dust or aggressive gases,
- strong electric or magnetic fields,
- explosive gases.

This list only applies to indoor use of the vehicles.

Spatial constraints can be, for example, limited room heights or load-bearing capacities of the floors.

If the above conditions do not show any abnormalities, it is still necessary to inspect the roadways, especially the ground. This is of fundamental importance for the safe and trouble-free operation of the AGV. The most influence can be exerted when the floor is newly constructed. Then the relevant standards and guidelines can be consulted, which can be found, for example, in VDI Guideline 2510-1.

A precise description of an "AGV-compliant" floor would go too far here. Generally speaking, it is defined by compliance with certain standards in the following criteria:

- Compressive strength of the pavement: The high surface pressure and the equally high shear forces are important.
- Friction: The coefficient of static friction should be between 0.6 and 0.8. If it is lower, proper emergency braking is not guaranteed; higher values will cause excessive wear on the wheels of the AGV.
- Flatness of the floor: this is all the more important the higher the demands on the accuracy of load transfer, e.g. when stacking into racks.
- Uphill and downhill stretches: Uphill stretches must be controllable by the vehicle drive, and downhill stretches involve risks in the event of emergency braking—the vehicle must neither tip over nor have longer braking distances. Sufficiently large transition radii (order of magnitude: 25 m) must also be ensured so that the AGVs do not touch down with the frame when driving up the slope and when leaving it, because the ground clearance of the vehicles is only a few centimetres for safety reasons. Five to seven percent gradients are normally no problem.
- Electrical discharge capacity: To avoid electrostatic charges, the floors should have a maximum earth discharge resistance of 1 MΩ. Often it is straight plastic floors that are both extremely smooth and extremely insulating.
- Cleanliness: The floors must be cleaned regularly during the operation of the AGV, making sure that the floors are completely dried after cleaning, because wet floors can lead to unsafe driving and braking manoeuvres.

The roadways on which the AGVs drive may normally be shared by other road users such as pedestrians, cyclists and forklift trucks. They must be visually marked as such. The minimum width of the drive track is calculated from the width of the AGV (including load), a margin of 50 cm on each side and, if necessary, a margin for meeting traffic of a further 40 cm. (Example: An AGV with a width of 1 m, which uses a route with two lanes—i.e. in two-way traffic— requires a regular track width of 2×1 m plus 2×0.5 m plus 0.4 m = 3.4 m).

Whether additional safety measures or facilities are required due to restricted traffic route widths must be clarified in the project phase with the Government Agency for Occupational Health and Safety and the responsible German Social Accident Insurance Institution (Note: Responsibilities might differ in other countries).

2.4.2   System-Specific Interfaces

The method used for **navigation and position determination** may require special markings (coloured stripes, metal bands, floor magnets, transponders, etc.) on or in the floor and/or also on the pillars and walls (reflectors, reflective marks, beacons, etc.).

An important system-specific interface is that to the stationary **load transfer devices**. These can be active or passive. One speaks of an active load transfer device if it has one or more electric drives. In this case, direct communication from the AGV or central control via the AGV master control system is required to control and, if necessary, synchronise the drives with the load handling device of the vehicle.

The transfer of the load is a safety-relevant situation that must be coordinated with the Health And Safety Authorities and the responsible Social Accident Insurance Institution. In any case, the endangerment of persons must be avoided. This is most easily achieved if the safety-relevant area is a closed area for which it is ensured that no persons are present—at least immediately before and during the load transfer.

Where this is not possible, special measures are common:

- Ground markings for marking hazardous areas,
- standstill preventers or baffles at the entrance to the load transfer station,
- stationary safety devices according to Table 2.8,
- optical and acoustic warning signals at the AGV's side,
- special sensors to detect people or other obstacles.

In many installations, the vehicle batteries have to be exchanged for charging. This can be done manually or in automatic battery changing systems. The AGV supplier must design and supply the interface for such automatic systems. If the batteries remain in the AGVs during charging, **automatic battery charging stations are** recommended, to which the vehicles are sent by the AGV control system for recharging. There is usually a central interface via LAN to the AGV master control system (Fig. 2.48).

Since the charging of batteries can lead to potentially harmful gassing, a number of regulations must be observed for the design of the corresponding premises, which can also be found in VDI 2510-1. Essentially, this is about sufficient ventilation so that the gas concentration does not reach inadmissible levels.

If the application is in hospital logistics, there is most likely an interface with **wheeled containers** (Fig. 2.49). These must be driven under, lifted and transported by an underrun AGV. When driving underneath, the exact position of the container must be detected so

Table 2.8   Stationary safety devices according to VDI 2510-1

Safety element/measure	Intended use
Parabolic mirror	Always useful at intersections that are difficult to see, especially when forklifts and other vehicles use the same traffic routes
Traffic lights	For intersections that cannot be seen. As a rule, the AGV requests the right of way and switches on so quickly that it does not have to stop
Barriers	The use of barriers can be useful if at certain times (shift change, end of work, lunchtime, etc.) a large flow of people crosses the roadway of the AGV
Rotating beacon	To warn people of approaching AGVs in unclear sections of the route, e.g. for
Light barrier/light curtain	the workers' safety and protection within warehouse aisles
Suspended flutter bands or chains	To make walking on surfaces more difficult
Further protective measures	Impact protection, deflector, shifting mat, floor marking, pendulum flap

Fig. 2.48 Typical battery charging station: Chargers mounted on the wall and copper charging contacts embedded in the floor. (Source: DS AUTOMOTION)

that it can be picked up safely. In addition—in most cases—a container code must be read which is located at the underside of the container.

From the point of view of automatic transport with underrun AGVs, the following requirements arise for the roller containers:

- Clear entry area at front: 660 mm (width) × 365 mm (height).
- Total weight (filled) max. up to 500 kg.
- Four rotatable castors with directional detent in longitudinal direction, of which two castor wheels with brake and one or two castor wheels as anti-static wheel.

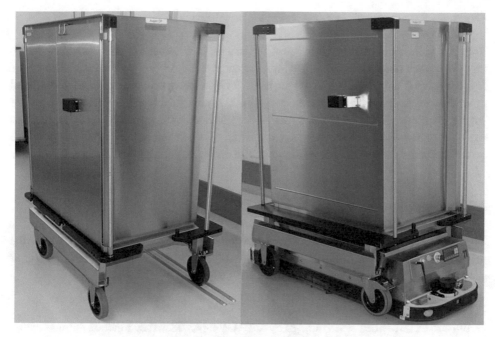

Fig. 2.49 A roller container in a hospital; on the right with underrun AGV. (Source: Hupfer, Coesfeld)

- The rotating container wheels must automatically lock in the parallel position to the vehicle when raised so that they do not collide with the side of the vehicle. This is usually done with an integrated spring.
- High-quality and smooth-running castors, approx. 180 mm diameter.
- Smooth-surfaced floor, load-bearing and tight against the container interior.
- A recess is to be provided in the centre of the floor for the possible accommodation of transponders, magnets or other ID tags.
- Container doors and flaps must "latch" to prevent unintentional opening while driving.
- Opening angle of the doors 270°; additionally, a detent for rear washing of the doors.
- CWS[24] requirements separately (stainless steel, temperature resistance etc.).

The last of the system-specific interfaces to be mentioned are devices that are required for **data transmission.** A data transmission takes place between

- the AGV master control system and the AGVs,
- the load transfer stations and the AGV master control system, but also to the vehicles,
- the AGV master control system and the automatic battery charging station,
- the vehicles among each other,

---

[24]CWS = Container Washing System.

• the AGV master control system and other peripheral equipment.

Today, data transmission between the AGV master control system and stationary equipment is mostly done via LAN. There are various possibilities for data transmission to the (mobile) AGV: Inductive, infrared light, radio. WLAN technology is the most widespread in modern systems, but there may be restrictions, requirements or even prohibitions of use by the plant operator if WLAN is already being used extensively in the company. In this case, alternative technologies such as narrowband radio, Bluetooth or ZigBee may have to be used. Ensure complete radio coverage of the entire AGVS deployment area.

2.4.3   Peripheral Interfaces

The third group of points of contact between the AGV and its environment concerns the peripheral equipment, i.e. the doors, gates, lifts, and any further automatic conveyor systems.

2.4.3.1 Doors and Gates

Doors and gates can be passed through by the AGVs, provided they function automatically. There are different ways to do this:

• The AGV master control system is able to control the door; there must be a LAN connection for this.
• The AGV communicates directly with the door control (e.g. via infrared data transmission or Bluetooth).
• The door is equipped with its own sensors and detects the approaching AGV. Suitable sensors are: contact loops in the floor, light barriers on the walls or motion detectors (based on radar or infrared light).

In any case, it should be ensured that the door is opened quickly enough so that the approaching AGVs do not have to delay or even stop and wait. After the AGV has passed through, the door can be closed again by exchanging signals.

If the doors are fire doors (also fire protection doors or fire section doors), they are normally operated in a similar way to those described above. Frequently, fire doors are also permanently open and close automatically in case of fire only—either locally controlled by connected fire detectors or centrally controlled by a fire alarm centre. This automatic control has a higher priority than the request of the AGV or the AGV master control system.

However, it must be ensured that the door does not close if an AGV is in the passageway and would become trapped. This would mean that the fire door cannot close completely and would therefore only fulfil its task to a limited extent. For this reason, fire doors or their controls are usually equipped with a time delay that gives the AGV about half a minute after the alarm is triggered to leave the door area.

2.4.3.2 Lifts

If the AGVs move in layouts across floors, elevators (also lifts, jacks or vertical conveyors) must be used. First of all, it is important to check whether these lifts are also used by people. If this is the case, the following points must be clarified in advance:

- Which persons are involved: trained staff, untrained staff, visitors, children, patients, etc.?
- What is the frequency of use by people and by the AGV?
- Are time windows conceivable in which passenger and AGV traffic can be separated?
- Can passenger transport be regulated with key switches so that only trained personnel operate the lift?

A possible mixed operation (AGV and persons) must be discussed in detail in any case and may also require special safety features in the AGV (recognition of persons within the lift's cabin and situation-related "making way" so that the persons can leave the lift).

The requirements to be placed on a lift from an AGVS' point of view are divided into mechanical and control requirements. The mechanical ones are:

- The clear length and the clear width of the cabin are calculated from the maximum length and width of an AGV (including load) plus 1000 mm. Thus, a retracted AGV has 500 mm of space on all sides of the car. This value can be reduced to a minimum of 200 mm with the special approval of the responsible Social Accident Insurance Institution.
- The clear door or cabin height must be the maximum height of an AGV (including lifted load) plus 100 mm.
- The clear door width results from the largest width of the AGV (including load) and an edge allowance of 200 mm on both sides.
- The required floor quality corresponds to that of the floor suitable for AGVs.
- The height difference between the cabin and the solid ground must not exceed ±5 mm. The cabin must not "sag" when driving on and must not "jerk up" when driving off— levelling may be necessary.
- The gap between the cabin floor and the fixed floor must not exceed 30 mm.
- If entry and exit are on the same side of the cabin, there must be sufficient space in front of the cabin for a waiting position of an AGV so that there is no mutual blocking of a vehicle ready to exit and a vehicle waiting for entry.

The lift interface is the connection between the AGV master control system and the lift controller, via which the AGV calls the lift and receives status messages. The following typical signals are sent from the AGV to the lift:

- Request automatic operation
- Go to the starting floor, i.e. the floor where the transport is to start
- Go to the destination floor, i.e. the floor where the transport is to end
- Cabin door must not close

The following signals typically are sent from the lift to the AGV:

- Automatic mode
- Lift arrived on starting floor, the door is open
- Lift arrived at destination floor, the door is open
- Lift status: Ready for operation
- Lift status: No fire alarm

The following boundary conditions must be observed:

- In AGV mode, the lift must not be operable by the floor call buttons or panels within the lift cabin.
- During entry and exit by the AGV, the lift must not make any movement, especially it must not close the door.
- The lift door must open automatically immediately after arrival on the destination floor.
- In the event of fire, the lift must function without an AGV master control system. In this case, the AGV master control system does not take over the control of the lift.
- It must be clarified with the local approval authorities whether AGV operation must be signalled visually (lamp or display) on the outside of the floors and/or inside the cabin.

2.4.3.3 Other Automatic Transport Systems

Other automatic transport systems may be operating in the immediate vicinity of the AGV. These can be crane systems or overhead electric monorails as well as track wagons or any further rail guided vehicles on the floor.

The cooperation of AGVs with crane systems must be checked with regard to the overlapping of the working/driving areas. Suspended crane hooks or loads are not detected by safety laser scanners for persons' protection and safety mounted on the AGV! If a load change from AGV to crane and/or vice versa is to be realised, the interfaces required for this must be coordinated between the parties involved on a project-specific basis.

If the AGVs have to cross rail tracks or other underfloor systems, a signal exchange must avoid any collisions. In addition, care must be taken to ensure that the requirements for the ground or coupling sensors and the braking distance are met.

2.4.4   Humans and AGV

And how does the interaction between the humans and the automated vehicles work? That depends on whether there is an interface at all and, if so, what kind of people have to be reckoned with. We can distinguish the following cases:

2.4.4.1 Segregated Areas

There is the rare case that the area of use of the AGVS is delimited from that of the people, i.e. the work area is sealed off. Mechanical boundaries, such as fences, or virtual boundaries, such as safety light curtains, keep the layout of the AGVS free from (unauthorised) people. The safety situation is comparable to the one we know from the working areas of industrial robots—as soon as a human, or even an arm, enters this area, there is an alarm and the system is stopped.

Now, there are hardly any such cases in indoor use. After all, it is precisely the "shared use" of the roadways by the AGVs that is at issue here. However, a separation of routes is not completely nonsensical here either. It may be possible to reserve some special lanes for the AGVs, perhaps in order to cover relatively long distances at "excessive" speed.

In outdoor areas, the creation of explicit AGV routes or entire areas is quite common. The special features of outdoor AGVs are explained in detail in Sect. 3.3 "Areas of application—outdoor use (outdoor AGVs)"; at this point it should suffice to point out that there are restrictions on the use of AGVs in outdoor areas due to the available safety technology. In any case, it is advisable to check carefully whether it is not possible to separate the occupied areas of people and AGVs. If this is the case, it is also possible to operate without extensive certified safety technology, as demonstrated by the large AGV application (almost 120 outdoor AGVs for the transport of ocean containers) at the HHLA[25] Container Terminal Altenwerder in Hamburg. In such cases, the question of the human interface is superfluous.

2.4.4.2 Staff Members

For the indoor area, it is a matter of course that the automated vehicles share the roadways with staff members. We are talking about intralogistics applications where there are certain ideas or requirements regarding workers—at least with regard to the subject matter discussed here. An employee is a healthy adult who moves responsibly in the company. He has been instructed in the handling of the AGV and has become accustomed to the automatic traffic participants.

---

[25] HHLA = Hamburger Hafen und Logistik AG.

In this way, the intralogistics applications are also controllable in terms of safety. The employees know how the AGVs move and where they are going. Experience teaches that within the first 2 weeks of use of the AGVS, the employees convince themselves that the safety sensor system also reacts to their person. During this time, there are a disproportionate number of stops, sometimes even with a loss of system performance. After that, there are hardly any problems in handling the AGV.

For the staff who are pedestrians, the AGVs show an easily predictable and pleasant behaviour:

- The normal driving speed of the vehicles is usually 1 m/s ($=$ 3.6 km/h), which corresponds to the usual walking speed of the staff. This avoids the situation where staff are frightened by vehicles approaching from behind.
- The AGVs always follow exactly the same drive tracks and do not usually drive around obstacles.[26] If employees are in the way of the AGVs on the roadway, the AGVs stop gently until they come to a complete stop and wait until the way is clear. Then the AGVs start up again independently.
- From experience, the AGVs become more likeable to the staff if they have names. There is a plant whose three vehicles are called Tick, Trick and Track. But the first names of the top bosses have also been given to AGVs.
- Electric vehicles drive almost silently, which is not always desirable. An alternative to the annoying continuous beeping of a warning device is to install a car radio on the AGV. Then the staff only has to agree on one station and the vehicles provide information and good humour and are also noticeable in time.

For forklift truck drivers, the situation is sometimes different. AGVs are often regarded as job killers. So here it is a question of providing information about the sense and purpose of the measures at an early stage, i.e. long before the introduction of an AGV system. Increasingly, however, companies are even automating because manual processes can no longer be staffed with suitable personnel (shortage of skilled workers).

The "cooperation" of forklift trucks and AGVs can work very well—provided that the forklift truck drivers "want to". Otherwise, collisions can occur in which the forks of the forklift "impale" the AGV and, for example, disable the expensive and sensitive safety laser scanner. In this case—with the appropriate consideration—a problem-free coexistence is possible because the forklift driver can observe and assess the movement behaviour of the AGV very well due to his elevated seating position. The traffic rule is that AGVs always have the right of way, and they also often drive more slowly and "behave" more

---

[26]Technically, automatic obstacle avoidance is possible today—whether it makes sense or is desired must be clarified on a case-by-case basis between the AGV supplier and the operator; of course, all applicable safety guidelines and regulations must also be complied with for obstacle avoidance.

carefully than the man-operated forklifts. AGVs therefore require special consideration, which is not always easy for forklift drivers.

2.4.4.3 Public Traffic

In some cases, there are other groups of people who come into contact with the AGVs in addition to instructed staff and trained personnel. An example of this is hospital logistics, where there is always public traffic in places. By far the largest proportion of roadways in the hospital will certainly be away from the patients' tracks—but not exclusively. Then an AGV suddenly encounters freshly operated patients with infusion stands or walkers, who are certainly restricted in their movements. Curious children or crawling toddlers cannot be ruled out either.

Situations like this pose a great challenge for the safety technology of the AGV. The approval of a laser safety scanner by the German Social Accident Insurance Institution loses its meaning here because we are moving outside the permitted area of application. Here, technical and organisational measures have to be taken to guarantee the greatest possible safety without the system performance of the AGVS suffering too much.

Areas of Application

3

After we got to know the technological standards in the last chapter, which are the basis for the many successful AGV realisations in recent years, we now want to look at the common areas of application of AGVs today, i.e. typical systems and techniques of the third era of AGVs. First we will categorise the applications according to processes and then look at examples of use from specific industries. A separate section is then dedicated to the special case of "outdoor AGVs", in which the special features of outdoor use with a focus on navigation and safety are described and explained with the help of some examples.

All the examples presented were realised in the period between the years 2000 and 2018 and should be understood as representative of their respective group—and not as a particularly successful project of the respective manufacturer.

3.1 Task-Related Aspects of AGV Use

The main areas of application for AGVs are in intralogistics, i.e. the organisation, control, implementation and optimisation of the internal flow of goods and materials, the flow of information and the handling of goods in trade and industry and in public institutions (definition according to VDMA[1]).

Some limitations are associated with this: For example, we do not consider the so-called people movers, i.e. automatic vehicles for passenger transport. This is also difficult at the moment: on the one hand, there are only very few applications and on the other hand, binding regulations and laws are largely lacking.

---

[1] VDMA = Association of German Mechanical and Plant Engineering.

© Springer Fachmedien Wiesbaden GmbH, part of Springer Nature 2023
G. Ullrich, T. Albrecht, *Automated Guided Vehicle Systems*,
https://doi.org/10.1007/978-3-658-35387-2_3

Many special applications[2] are also left out: Applications in space travel, in or under water, in military technology, facade and floor cleaning, mobile information systems for visitors to museums, exhibitions, shopping centres, etc. as well as walking or climbing machines.

So we limit ourselves to the transport of material, especially in intralogistics.

3.1.1   The AGV in Production and Services

To begin with, let's take a closer look at the tasks of intralogistics, because this is the environment in which the classic AGV operates.

The movement of goods (general cargo, liquids, merchandise, materials, supplies, etc.) takes place in different areas within a company or company premises, between physically separated companies or parts of companies, between companies and consumers.

The organisation, implementation and optimisation of these flows of goods and materials within a company in industry, trade or a public institution are referred to as intralogistics. Essential aspects of this comprehensive subject area are

- the processes of handling goods and materials, in particular incoming and outgoing goods, warehousing and order picking, transport and the handover and provision of the same;
- the information flows, i.e. the communication of stock and movement overviews, the order situation, throughput times and availability forecasts, the presentation of data to support the tracking, monitoring and, if necessary, decision-making of measures, and also the selection and use of means for data communication;
- the use of means of transport (hoists, continuous conveyors, industrial trucks, etc.) as well as monitoring and control elements (sensors/actuators);
- and finally the use of techniques for active/passive security, data management, goods, merchandise and material recognition/identification, image processing, goods handling (i.e. staging, sorting, picking, palletising, packing).

In the vast majority of cases, transport processes do not add value, but may cause considerable expense. On the other hand, since transports are necessary for internal processes, there is both the challenge and the opportunity to optimise them! In the interdependency of the means of production, the selection and design of the transport systems influences the efficiency of the production process and thus its earnings potential.

The **production area** is characterised by the process chain from goods in to goods out. Influenced by the order situation, purchasing, scheduling, production management and administration continuously design various elements of this process chain, i.e. essentially

---

[2] A first automated car parking system with AGV was realised by Serva Transport and Fraunhofer IML, Dortmund, at Düsseldorf Airport in 2013. Source: Hebezeuge Fördermittel, Berlin, issue 53 (2013), p. 6.

- the build-up and reduction of stocks and the necessary handling of goods, merchandise and materials (incoming and outgoing goods, material storage),
- the set-up times and throughput times, taking into account the balancing of over- and undercapacities as well as the delivery targets of the service recipients,
- the definition or modification of order priorities; and
- the optimisation of batch sizes.

These tasks require permanent control, monitoring, supervision and mostly adaptation to the constantly changing situation. In order to achieve the greatest possible scope for efficient performance of these tasks, careful planning (simulation if necessary) and a balanced use of suitable means of transport are therefore essential in addition to work-step-oriented production planning.

The same applies to applications in the **service sector**. If we understand the "production area" as the area that makes its services available to the recipient, then we see comparable tasks in the process chain, even if those responsible may have other job titles.

In the business management area of the company, the selection of the means of production mainly has an impact on

- the financial planning and use of funds, and
- the capacity and utilisation analyses and planning, concerning both technical means and, above all, human resources.

The technical and business management strives to constantly optimise the available financial and human resources in the area of conflict between the operationally necessary performance requirements and the resources needed for this. This requires suitably defined and recorded operational data and key figures, such as stock turnover times, throughput times with idle times, utilisation of production resources and the like.

This should suffice at this point to classify AGV systems in intralogistics, so that we can go into the role of the AGV systems in more concrete terms in the following.

3.1.2 AGV System as an Organisational Tool

Automated guided vehicles (AGVs) are often equated with AGV systems (AGVS). The discussion quickly turns to the different types of vehicles or other specific topics:

- Which type of AGV, e.g. forklift or underrun AGV, is preferable?
- Which is the preferred navigation method (e.g. laser triangulation or magnet navigation)?
- Which concept for personal safety should be used?

Of course, the AGVs are important components of an AGVS, but only components. If we want to be correct, we have to look at the overall AGV system, which according to VDI

2510^3 consists of the vehicles, the master control system and the floor system. This guideline lists essential global properties of the AGV (Fig. 3.1).

It must be emphasised here that an AGVS as an organisational tool has a far-reaching and lasting effect on intralogistics. At first, the order required as a prerequisite for AGV operation seems annoying. But then it becomes clear that this order is also the consequence of an AGV, so that there is an inbuilt opportunity to optimise the processes further and further in the sense of continuous improvement.

For example, when it comes to automating a typical forklift operation, i.e. intralogistics with manually operated industrial trucks, by means of AGVs, the operator may initially mourn the supposed advantages of the forklift: the high system performance that can be called up at short notice and high flexibility with regard to the task. But if he then takes a closer look, he realises that an AGV also has a high system performance, namely exactly the one that was "set" during planning; and quite naturally as a continuous performance with an extremely high availability.

The high flexibility of the forklift is only needed when the task has not been optimally structured (task of planning) or, in rare cases, cannot be structured. In most cases, however, the processes have sufficient potential for optimisation so that the processes can be organised in such a way that an AGV can be used. The repeatedly underestimated advantage of AGVs then lies in the fact that the created order is maintained in the long run because it has to be maintained! Examples of this are the clear definition of roadways (for driving) and parking spaces (for all kind of goods, parts, vehicles etc.)

Until a few years ago (approx. 2015), automatic vehicles could not drive around obstacles that were in the way, i.e. they stopped in front of them; obstacles such as a group of employees having a discussion or a pallet that had been placed down by employees "for a moment". This is also perfectly acceptable, because in a well-organised, automated operation, neither staff meetings nor occasional misplaced pallets on the roadways should interfere!

New navigation methods in combination with sensor technology + complex software aboard the vehicles make it now possible for an AGV to react to such a disturbance by driving around an obstacle independently, but carefully and slowly—provided there is sufficient space available. In individual cases, this is often seen as a positive feature, but when obstacle avoidance becomes a permanent condition, the transport performance of the AGV drops significantly.

By observing the actual state and adapting simple rules in the AGV master control system, it is then possible to maintain positive changes or reverse negative ones. The ongoing changes in the processes/product range/quantities etc. can thus be directly taken into account with adapted intralogistics. In this way, an AGV can optimise logistics processes with simple rules and grow with the requirements.

---

[3] VDI 2510 "Automated Guided Vehicles (AGV)", VDI 10/2005, Beuth-Verlag, Berlin.

Fig. 3.1 An AGV links different processes in paper roll handling. (Schematic; Source: Rocla)

3.1.3   Arguments in Favour the Use of AGVs

At this point we would like to summarise the advantages of the AGV. They will certainly reappear in isolated or modified form in other parts of this guide, but here we want to present the arguments in a bundle. The focus is not on economic efficiency, which must be given in any case, but on the technical and organisational arguments:

- Organised flow of materials and information; thereby increasing the transparency of internal logistics processes and boosting productivity
- Punctual and calculable transport operations at all times
- Minimisation of scare stocks and waiting stocks in the production area
- Reduction of personnel commitment in transport and thus reduction of personnel costs (especially in multi-shift operation)
- Minimisation of transport damage and incorrect deliveries; thus avoidance of consequential costs
- High availability and reliability
- Improving the working environment; safe and more pleasant working conditions through orderly processes, clean and quiet transport operations.
- Positive internal effect on the workforce
- Positive external impact within the corporate group (securing the company location)
- Positive external image towards customers
- Automatic load transfer with high precision
- Minor infrastructure measures required
- Easy realisation of intersections, merges and branches
- Multiple use of the transport level (= floor) possible
- Possibility of using a substitute means of transport (e.g. forklift truck)
- Suitable for both low and high room heights

- High transparency of the transportation process
- Usually no additional space for drive tracks required
- Use of existing roadways
- Indoor and outdoor use possible
- A wide range of additional functions can be realised:

    Ordering/sorting, deciding, collecting data, forwarding data, weighing transport goods, organising processes, managing warehouses, managing storage spaces, recognising loads, mastering different layouts, finding pallets, loading trucks, intelligent safety, behaving intelligently and according to the situation (special behaviour in case of fire, different operation modes), additional activities during low-operating times (e.g. night-time transfer), intelligent battery charging strategies, mobile robot, order picking functions, etc.

The traceability of the logistic processes is extremely modern. All product movements are reliably completed and logged. This creates a complete process history, which is useful and necessary for internal audits, but also in terms of product liability. In summary, it can be said that an AGV is a powerful tool for handling and, above all, optimising intralogistical processes. The listed features and advantages apply independently of and across all sectors! The examples of applications presented in Sect. 3.2 may show industry-specific vehicle solutions, but these are essentially due to the goods being transported or the loading equipment typical of the industry.

3.1.4 AGVs in Taxi Operations

Usually, AGVs are divided into flow line and taxi operation with regard to the forms of use. AGVs that operate as assembly platforms and that are scheduled by assembly lines operate in flow line mode, which will be the topic of the next section.

In taxi operations, the stations (similar to stops) are also called sources and sinks. Material transport begins at a source and ends at a sink. Of course, each station can also be a source and a sink at the same time (Fig. 3.2).

Vehicles that drive in a network of sources and sinks and link many individual positions freely and flexibly are part of a taxi system. Such an AGV can be compared very well to a taxi company in a city.

However, not only efficient vehicles are important for a taxi system. Of utmost importance is the AGV master control system (analogy: taxi control centre), where all information comes together and is optimally evaluated. This is where the potential for optimisation ultimately lies. To remain in the image of the taxi company in a city: To set up a successful taxi company, it is not enough to buy several cars. A taxi control centre is needed that receives all driving orders and is always well informed (current position of the individual taxis, current traffic situation in the city. . .).

Fig. 3.2 Principle sketch of taxi operation according to VDI 2710-1 (VDI 2710 Sheet 1: Holistic planning of AGVs—Decision criteria for the selection of a conveyor system; VDI 08/2007, Beuth-Verlag, Berlin)

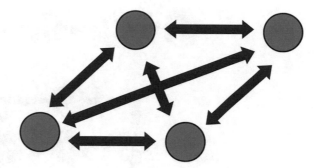

In the taxi control centre, the thinking happens so that transport orders can be assigned to different types of vehicles and all orders are carried out on time. In the process, a wealth of constraints are taken into account, such as priorities, time-limited layout restrictions (construction sites), daily schedules, etc.

The classic transport order for an AGV in taxi operation is: "Pick up from source X and bring to sink Y". This order is managed in the AGV master control system, i.e. it receives it and selects the most suitable vehicle to carry it out. Similarly, a passenger orders a taxi to take him from the hotel to the city centre, for example, by calling the taxi control centre. The taxi centre then ensures that a suitable taxi carries out this "transport order".

In addition to the above-mentioned functions of order management and vehicle dispatching, the classic AGV master control system typically has other tasks, which have already been explained in Sect. 2.2.

Such taxi systems are usually used to supply and dispose of machines in production or to link production/production areas with warehouses and goods out.

3.1.5 Assembly Line Operation and the Focus on Series Assembly

The first AGVs could only be used in assembly line operation. They lacked today's possibilities to realise complex layouts and to control the whole thing according to requirements. Today, flow line systems are mostly used in assembly systems (Fig. 3.3).

When designing assembly systems, there are many ways to realise the transports. We are talking here about the assembly of units in series, such as the classic engine assembly in the automotive industry. Related assemblies in the automotive industry concern cylinder heads, gearboxes, drive sets, steering systems, axles, doors and cockpits.

There are also comparable assemblies in many other sectors, such as in the electrical/electronics industry, in the white and brown goods sector or in mechanical engineering. Which transport system is used is subject to different criteria that are of a technical, economic or even (company) philosophical nature.

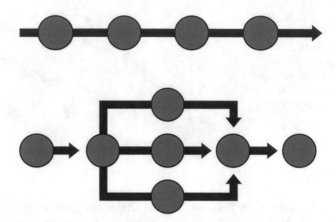

Fig. 3.3 Principle sketches of flow line operation, linear and branched (according to VDI 2710-1)

3.1.5.1 Tasks in Series Assembly

In such assembly areas, there are three main tasks for transport systems:

First of all, there is the main assembly line, in which the product to be assembled (e.g. the car engine, a gearbox or a car cockpit) has to be transported from the starting point of assembly to the end point. Between these two points the actual assembly takes place (application assembly line) (Figs. 3.4 and 3.5).

The second task concerns the picking of parts for assembly. Since the variety of types is usually too large for all necessary parts to be stocked directly at the assembly line, certain parts are assembled according to type in a separate picking zone. A transport system is also required to transport the order-picking containers, e.g. baskets, small load carriers (totes, bins), pallets or wire mesh boxes (order-picking application) (Fig. 3.6).

Last but not least, the picking totes must be transported from the picking zone to the assembly line. This transport does not necessarily have to be carried out with the same transport system that is used in the picking zone (application transport).

In practice, the following transport systems can be found in series assembly in addition to the AGV:

Rigid assembly line (RAL): This name is used here to refer to different technical designs of how the workpiece carriers are transported through the assembly line at a constant feed rate or in an intermittent clocked manner: e.g. with chain conveyors, roller conveyors or via underfloor chain-pulled conveyors.

Electric Monorail System (EMS): Electrically individually driven trolleys move on a rail that is either suspended from the hall ceiling or attached to an elevated steel structure. The trolleys are fitted with hangers that hold the goods to be transported.

Industrial Truck: This term covers manually operated vehicles, e.g. forklift trucks or tractors. We also include hand trucks in this group (Fig. 3.7).

In principle, all the transport systems mentioned can be used for the tasks. However, a first consideration of the economic efficiency leads to the following table of suitability (Table 3.1).

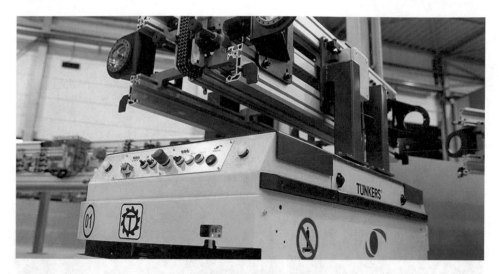

Fig. 3.4 Assembly of aggregates on the AGV. (Source: Tünkers Maschinenbau)

Fig. 3.5 Cockpit assembly on the AGV. (Source: BÄR Automation, BMW plant Oxford)

Fig. 3.6 Automatic pallet trucks in use in order picking. (Source: Rocla)

Rigid assembly lines are only suitable for the assembly line because they are too rigid, inflexible, slow and expensive for order picking and pure transport. In addition, paths are obstructed and passages blocked.

The manual industrial trucks make sense for pure transport, as the labour costs would be too high in the slow assembly and lengthy picking. In order picking, the driver would have to pick the parts himself, but this does not seem to make much sense in view of the cumbersome climbing up and down onto the driver's seat. The hand trolley is the inexpensive alternative here.

The Electric Monorail System holds a special position. Here, everything depends on the space available and the hall/ceiling conditions. But in most cases there will be more flexible and cheaper solutions. So the EMS can only be compared with its competitors to a limited extent.

Fig. 3.7 A simple trailer-pulling AGV (automatic tugger train) tows the transport trolleys, which are loaded with installation parts on a commission basis, to the assembly line. (Source: DS AUTOMOTION)

Table 3.1 Suitability table of the transportation techniques for the tasks in series assembly

	Assembly line	Order picking	Transport
RAL	+	O	O
AGV system	+	+	+
EMS	O	O	O
Industrial truck	–	O	+

Explanation of the symbols: – not suitable; O conditionally suitable; + very suitable

For the reasons mentioned above, only three pairings will be discussed below:

For the main assembly line:	AGVS vs. RAL
For transport:	AGVS vs. Industrial Truck
For commissioning:	AGVS vs. Industrial Truck (actually only handcart)

3.1.5.2 AGV or EMS in the Assembly Line?

Here, first of all, many details depend on the layout of the line: The simpler the layout, the more suitable the EMS. This is because the realisation of branches and island solutions is easily possible with the AGV only, and it does not cause additional costs. The layout also becomes more demanding when there are clocked production areas as well as flow assembly areas. Complex layouts also mean extreme accessibility restrictions with the EMS. Areas are obstructed, paths are blocked and the ergonomics of assembly suffer.

If you limit yourself to a "simple oval", the EMS has the advantage of being able to easily set constant feed rates. The rigid continuous assembly does not need an AGV! In addition, the technology of the EMS is simpler compared to the AGV, and therefore more robust and reliable.

But often the "simple oval" is just an unrealistic wishful thinking: What does the inspection concept in assembly look like, for example? Is there a 100% inspection or is it only a statistical inspection? Do individual workpiece carriers have to go to separate test stations from time to time?

The same question arises for reworking: What shall happen when defects are discovered? In most cases, it makes no sense to continue to complete defective workpieces, i.e. to feed them through with the other workpieces regardless of the defect. So you need your own production islands again.

The most important argument in favour of the AGV is the layout flexibility: If one wants to change the layout during the operation of the line, it becomes difficult and expensive with the EMS. Changes to the line can occur when the assembly contents change (e.g. with new products), because the quantities produced change or simply because optimisations in the process are to be implemented after a certain experience. This is where the AGV comes into its own.

In addition to the layout of the assembly line, another criterion is of crucial importance: Are there automatic stations in the line that require high-precision positioning and the absorption of high forces and moments? An example of such lines is the engine installation into the fuselage. This is no problem for the EMS: The workpiece carrier is fixed with high precision by mechanical guide rails. Stations like this can also be done with the AGV, but with considerably more effort. So ultimately it depends on the number of such automatic stations.

The smaller space requirement is often cited as an advantage of the EMS over the AGV. This argument is correct insofar as the space requirement of the assembly line can be minimised with the EMS—if it has to be (because only a certain area is available) or is absolutely desired (company philosophy).

But this argument must not be considered in isolation. The less space available in assembly, the less material to be assembled can be stocked directly on site. This means that picked parts have to be delivered to the assembly line just-in-time. This in turn means that extensive order picking has to be set up elsewhere. The space requirement in total, i.e. in assembly and order picking, will be almost the same. This does not take into account the space required for transport from order picking to assembly or the shared use of existing roadways.

3.1.5.3 AGV or Forklift Truck for Order Picking and Transportation?

In order picking and pure transport, it is a matter of comparing the manually operated industrial truck (or handcart) with the AGV, i.e. a manual with an automated transport system. Because there are a number of differences in principle that are independent of the specific task, these differences are presented beforehand.

There are generally two groups of arguments in favour of automation. The first group revolves around the quality of the transports. An AGV system prevents all kinds of transport damage, both damage to the loading equipment or goods being transported and damage to the stationary equipment such as columns, walls, racks, shelves, gates, etc. The first group is about the quality of the transports.

Ultimately, the AGV system should be seen as an organisational tool, as explained earlier. It ensures an optimal flow of material and information and thus more transparency. In addition, there are no more wrong deliveries: automation ensures absolutely reliable transports.

In general, it is certainly the case that with the development of computer technology—and the associated control and sensor technology—the market for automated material flow systems and thus also for AGVs is growing. Especially with regard to product and producer liability, which drives many industries to document every single process step of their production, this means the necessity of automation.

The second group of arguments concerns the ideal advantages of AGVs over manual conveying technologies: State-of-the-art logistics signals a technological advantage both internally and externally and thus has an image and motivational effect that should not be underestimated.

3.1.5.4 AGVs or Simple Handcarts in Order Picking?
But what does this mean for order picking? The ostensible advantage of handcarts is clear: the investment and operating costs are negligible compared to AGVs. Those for whom the system's performance is sufficient are well served by the manual variant—as long as they don't think outside the box when it comes to order picking! After all, the totes or baskets of goods that have been filled on the handcart still have to get to the assembly line somehow—and certainly not by a worker pulling or pushing the handcart!

The AGV has further quality advantages in order picking: Forced sequences can be programmed, i.e. fixed picking locations and destinations can be specified. A continuous documentation of the processes takes place without additional effort. In addition, mixed continuous flow and intermittent cycled picking areas can only be realised with an AGV.

3.1.5.5 AGV or Forklift Truck for Pure Transportation?
When the order-picking totes or baskets of goods have to be transported to the assembly lines, there is the classic competition between forklift trucks (or tractors) and automated guided vehicles. In addition to the above-mentioned general differences between manual and automated transportation technology, the comparison here focuses on the following points:

If one compares the costs of the systems, and not just the initial investment, but in terms of TCO,[4] then the initially higher investment for the AGV is put into perspective by the

---

[4]TCO = Total Cost Of Ownership.

lower operating costs and, in particular, by the lower maintenance costs, because the automation technology is much gentler with the technology: due to the even and gentle driving style, the wear and tear on tyres, batteries, drives, etc. is significantly reduced. Moreover, it is an open secret that automated guided vehicles are designed for continuous use of far more than 10 years, in contrast to lifetimes of approx. 3–4 years for forklift trucks.

In order to complete the profitability comparison, the personnel costs must of course be taken into account, which are of decisive importance for forklift trucks and tractors. In two-shift operation, three drivers per vehicle have to be reckoned with, in three-shift operation it is even about 4.5. With annual full costs of about 40,000 € per driver, any supposed investment advantage of the manual systems begins to waver.

The comparison of the industrial truck with the AGV shows further ideal advantages for the AGV, e.g. in terms of availability and continuity. The AGV works unspectacularly, without interruption and without any hectic. It also works safely and without accidents. It creates order and cleanliness; stress is reduced and a pleasant working atmosphere is created.

3.1.5.6 Conclusions for Series Assembly

The AGV is a serious alternative for all three tasks discussed here. Since series assemblies are usually not designed for a short period of time, but for periods of 5 or more years (of course with regular layout adjustments), there is much to be said for the AGV because of the low operating costs and the low effort required for changes.

And consistently for all three tasks: The AGV in assembly, in order picking and for transport in between—the universal variant without changing the transport system!

And: Why should you put add-on parts in a shopping basket, on a tray or shelf, only for it to be picked up again a short time later and mounted on the workpiece, which would be nonsensical from an MTM[5] point of view. It is better to send the workpiece on the AGV through the order picking and to carry out pre-assembly already there: Complicated add-on parts are not placed next to the workpiece, but are already fixed to the workpiece, so that the classic division of assembly and order picking is broken up.

Although the AGV requires more space in assembly than the EMS, taking into account the space requirements in order picking, the AGV provides a holistically improved solution when planned holistically!

With an additional option that only the AGV offers: Planning with its extensive MTM analyses reaches its limits in everyday life. If AGV assembly lines are already planned with corresponding stand-alone solutions or alternating workplaces, the foremen, supervisors and employees can carry out their own optimisations on site without much effort. This increases the self-responsibility of the assembly workers and boosts their motivation—all in the spirit of continuous productivity improvement.

---

[5]MTM = Methods-Time Measurement, which can be translated as method time measurement.

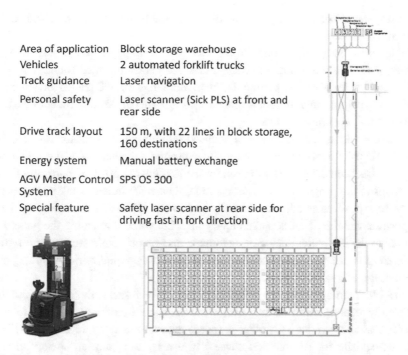

Area of application	Block storage warehouse
Vehicles	2 automated forklift trucks
Track guidance	Laser navigation
Personal safety	Laser scanner (Sick PLS) at front and rear side
Drive track layout	150 m, with 22 lines in block storage, 160 destinations
Energy system	Manual battery exchange
AGV Master Control System	SPS OS 300
Special feature	Safety laser scanner at rear side for driving fast in fork direction

Fig. 3.8 Example of a floor-level block storage system. (Source: E&K Automation)

3.1.6 Storage and Picking

Two of the main tasks in intralogistics are *storage* and *order picking*. In the following, we want to focus exclusively on the topic of block storage, because a free storage area seems to be fundamentally predestined for automated guided vehicles.

3.1.6.1 Floor Level Block Storage

Simple floor-level block storage systems operated by AGVs have been around for many years in various forms. Starting with individual buffer rows for the delivery or pick-up of pallets provided in production or in the warehouse, up to buffer rows filling areas (Fig. 3.8).

A typical example from practice is like this: production is running over three shifts and goods out is running over two shifts only. The AGV buffers the running production overnight in a floor-level block storage at the goods out area. The AGV is thus used for the automatic transport and the block storage.

A typical task for AGVs can also be the provision of pallets for truck loading.

3.1.6.2 High Block Storage

Block storage systems are often multi-level, i.e. several pallets are stacked on top of each other. The automation of such a "high block storage" is naturally demanding and requires high-quality and more or less uniformly packed pallets. With powerful vehicles and an intelligent system control, highly flexible block storage facilities with thousands of storage spaces can be realised.

This is particularly interesting in the beverage and paper industries, where large quantities of relatively uniform and easily stackable goods are often found. One example is the Radeberger brewery in Dortmund. There, more than 10,000 pallets are automatically stacked on top of each other in four levels on a storage area of 80×80 m, with the top layer of beer crates at a height of 6.6 m.

This high stacking can no longer be realised reliably without sensor support. For this purpose, a 3D pallet recognition system with subsequent evaluation of the actual position of the pallet has been developed and used for the first time in 2011 by company E&K. The AGV moves to the programmed target position, then a 3D image is taken and evaluated: Then the fork position is adjusted laterally and in the lift height according to the actual pallet position and the pallet is picked up cleanly, i.e. without touching the pallet feet.

Thanks to the 3D pallet detection, automatic and reproducible storage and retrieval is possible, even at great heights. Individually tailored storage and retrieval strategies allow for optimal use of available space.

Since a block storage system completely dispenses with racks, the user has the freedom to completely reconfigure his warehouse, to move or rotate storage rows or to temporarily block off areas and use them for other purposes without much technical effort (Fig. 3.9).

The prerequisite for reliable and successful use is, as in any automatic operation, a supervised quality of the pallets and also a sufficiently levelled floor in accordance with the Technical Conditions for AGVs.

The division of the storage blocks depends on the shape of the hall, the storage and retrieval points, the number of items, the storage strategy and the transport capacity. There is an optimal solution for every requirement. When planning an AGV-operated block storage system, it is essential to compare the special requirements with the shape of the hall and the possible storage structures.

3.2   Sector-Related Aspects and Examples

In contrast to the previous section, in which we approached the areas of application from the transport task, the examples described below are intended to highlight aspects of the respective AGV application that are typical for the industry. The example character is independent of the time of realisation, i.e. the transferability of the examples to current situations in the company of the reader of this guide should be given at any time. The third era of AGVs is characterised by the fact that more and more sectors of application are opening up to AGVs. This sounds good at first, because new AGV providers are opening up new markets step by step. However, one must not forget that this third epoch came virtually out of nowhere and that the sudden market collapse at the beginning of the 1990s had to be compensated for. The new beginning in the mid-1990s happened with a new bidding landscape, which was then for the first time characterised by medium-sized

3D pallet detection allows compact block storage warehouses, especially in beverage industry. The drive track layout can easily be changed by the end user.

Application	More than 10,000 pallet locations in a hall 80 x 80 m
Vehicles	10 automated counterbalance forklift trucks, lifting height 5 m
Track guidance	Magnet navigation (point sequence)
Personal safety	Safety laser scanner (Sick PLS) at front and rear side
Energy system	Manual battery exchange to enable 24 hours of operation
AGV Master Control System	SPS OS 820
Communication	WiFi
Special feature	3D pallet detection

Fig. 3.9 Features of an AGV with 3D pallet detection in block storage. (Source: E&K Automation)

companies, most of which had between 40 and 80 employees. It was not until about 20 years later that large companies entered the AGV business again.

While during the second era of AGVs a few large suppliers concentrated on a single sector of application, namely the automotive industry, during the third era of AGVs many small and medium-sized companies have to deal with a multitude of very different user industries. This is what made the AGV business so versatile.

The AGV business is extremely exciting because companies can use all their engineering skills to sell a sophisticated product to a wide variety of markets. But at the same time, of course, it is also difficult because the sales process is usually largely passive. So what does passive or active sales process mean in the project business that prevails here?

A passive sales process is characterised by the fact that the provider only responds to incoming project enquiries due to its limited resources and the infinite variety of sectors. An RFP[6] received by the provider by mail, fax or e-mail is processed. The advantage is that all resources are used for processing real enquiries and costs for active-strategic doing can be saved.

An active sales process means that the AGV provider focuses on a few target industries and actively markets there. This does not mean that enquiries from other sectors may not be processed. It only means that the provider knows the selected target industries inside out and, as a competent thinking and compassionate system partner, can hope for better chances in the awarding of projects.

---

[6]RFP—Request For Proposal.

How actively or passively the individual providers conduct their business is up to them. In any case, the AGV business has always been demanding. In the third era, there were a lot of small and medium-sized companies trying to win every single AGV project, and the price pressure was high. The sales teams prepared too many offers that did not lead to an order after all. Prices were usually extremely negotiated, so profit margins in the projects were low, especially since many projects held unforeseen risks due to their complexity. In the fourth epoch, the previous players have been joined by both some new small ones and some larger ones (corporations), and there is a highly increased market demand on AGV systems. Now, the big challenges are not the low prices, but the extremely high utilisation of the AGV suppliers.

But let's return to the user industries and the growing opportunities that are emerging worldwide. The technology offers a modular system with which reliable and efficient AGV solutions are feasible. The quality requirements in the target industries are increasing and so are the personnel costs—so far a positive prognosis as far as the markets are concerned. In the following, we list examples of industries and projects and provide industry-specific information.

3.2.1 Automotive and Supplier Industry

At first it seems paradoxical to start with the automotive industry, which had completely abandoned the use of AGVs around 1990. But after a few years of abstinence, the first AGV projects found their way back into the car manufacturing plants at the end of the 1990s. Today, the automotive industry is calling for extremely flexible structures in its production facilities and is by far the largest user industry for AGVs again. Since around 2015, even suppliers have discovered AGVs, as they were extremely rare AGV customers in the past. Here are a few examples to show that there are very different applications, from the emphatically simple, i.e. following the Japanese KAIZEN idea, to functional, technically demanding but reasonable, to unusual applications.

3.2.1.1 AGVS in the Transparent Manufacture Dresden (Volkswagen)

In the "Transparent Manufacture" in Dresden, Volkswagen AG assembled the then new luxury class model "Phaeton" from 2002 to 2016. An transport system with 56 automatic vehicles took over the material supply to the assembly lines. The AGV system was supplied jointly by FROG (vehicle control and navigation, AGV master control system) and AFT[7] (mechanics) (Fig. 3.10).

The "Transparent Manufacture" was never an ordinary automobile plant. The high demands placed on the new product (first the Phaeton model, today the E-Golf) were also placed on the production facility. The facilities and equipment present themselves

---

[7] AFT—Automatisierungs- und Fördertechnik GmbH & Co. KG, D-Schopfheim.

Fig. 3.10 View of the assembly line. A shingle conveyor embedded in the floor winds its way through the assembly line. (Source: Volkswagen)

bright and friendly, the space is generous and the floor is covered with valuable maple parquet. The organisation of work was also unique: the emphasis on handicraft "manufacture-like" activities as an antithesis to purely performance-based assembly lines, combined with the most sophisticated innovative technology, was intended to match the image of the "Phaeton" and ensure its high quality standard.

Production is spread over three levels. The actual assembly takes place on the two upper assembly levels. The car body is placed on an assembly platform that is part of the shingle conveyor that fits flush into the hall floor and moves through the assembly cycles at a constant speed. The car body is then transferred to an electric monorail system (EMS) for suspended assembly. During the overhead assembly, the "wedding", i.e. the joining of the car's body and the drive set, takes place, whereby the drive set is brought in by an automated guided vehicle. Afterwards, the body is placed back on a pushing platform, the so-called shed, for completion and quality control.

In the basement, the logistics level, the material to be assembled is provided and picked. The AGVS takes over the supply of the assembly lines with this material and thus an important logistical function. To switch between the levels, the automated vehicles use lifts.

One special feature concerns navigation: on the assembly levels of the "Transparent Manufacture", maple parquet flooring of the same size was laid instead of the usual screed. These were each fitted with four permanent magnets before installation. The magnets, which have a diameter of 8 mm and a height of 5 mm, were pressed into blind holes on the underside of the boards and sealed with a filling compound. As a result of this measure, there is now a continuous magnetic grid on the assembly levels for completely flexible track design.

The AGVS has the basic task of supplying the assembly lines. The following types of goods must be delivered:

1. Shopping baskets on the shed and for suspended mounting,
2. control panels (cockpits),
3. wiring harnesses,
4. engine and gear box with chassis and "wedding" execution,
5. doors plus additional baskets.

The different types of goods require two different types of vehicles. The small vehicles (for positions 1–3) offer a load capacity of 800 kg and a differential drive, i.e. two separately driven, non-steered wheels in the middle of the vehicle and one freely rotating support wheel in the middle at the front and rear. It can thus drive forward and backward with equal precision, realise very tight curve radii and turn on the spot.

The large vehicles transport the engine with chassis and the doors (positions 4 and 5). Its load capacity is 2500 kg and they are 1 m longer than the small ones. The large AGV has a 4-wheeler kinematic—one driven and steered wheel each at the front left and rear right, as well as two castor wheels at the front right and rear left. With this chassis, the vehicle can move omni-directionally, i.e. in addition to straights and curves, diagonal and transverse rides are also possible. Especially the transverse ride is required at the wedding station, where the power unit with the chassis (on the AGV) is joined to the body (at the EMS) (Figs. 3.11 and 3.12).

The transport of the shopping baskets to the shingle conveyor is completely new and places the highest demands on the AGV and AGV control. After order picking the AGVs pick up a shopping basket on the logistics level by driving under it and lifting it slightly. Then they bring it to the assembly level with the help of the lift. There they have to place the shopping basket, which is filled with assembly material for a certain car, on a certain shingle. To do this, the AGV has to drive up from the fixed hall floor onto the slowly moving shingle conveyor. The AGV master control system instructs the selected AGV to transport a shopping cart to the waiting point close to the shingle conveyor and wait there. The position of the passing shingle is constantly reported by the shingle conveyor controller (a PLC) to the master control system. With the help of a shopping basket identification, a check of the correct shopping basket is guaranteed.

As soon as the stop position on the shingle is opposite the waiting position of the AGV, the AGV receives the order to drive onto the shingle conveyor with a lead time that depends

Fig. 3.11 The wedding station: A body in the EMS chassis approaches from above, an AGV with the drive train is waiting in the foreground below, and ana AGV with a shopping basket passes in the background. (Source: Volkswagen)

on the shingle speed. After driving onto the shingle, the positioning accuracy is approx. 10 mm, but this is immediately corrected by the magnetic grid that also exists on the shingles. Leaving the shingle conveyor is done in an analogous way.

3.2.1.2 Production of the BMW 3 Series at the Leipzig Plant

In 2005, the new BMW plant in Leipzig started production of the 3 Series (E90). In the area of parts supply, an AGV system took over extensive logistics functions for the first time in the history of passenger car production (Fig. 3.13).

The following standard processes have been defined for the supply of parts to the assembly department:

1. Direct delivery by truck: Large parts with low complexity (e.g. floor mats or boot linings) are delivered by truck promptly and in the immediate vicinity of the installation site.
2. Module delivery by EMS: Large and complex assemblies (e.g. cockpit) are assembled directly on the factory premises by external suppliers or BMW workers.
3. Stocked goods by AGV: The majority of parts are stored in a supply centre, picked and transported by AGV to the respective assembly locations.

Fig. 3.12 Once again the wedding station: in front an empty AGV moves away, behind it the "wedding": drive train from below joins body from above. (Source: Volkswagen)

Fig. 3.13 Underrun AGV with small trolleys for pallet cages. (Source: DS AUTOMOTION)

Fig. 3.14 AGV with oversized trolleys for holding large containers. (Source: DS AUTOMOTION)

The entire drive track layout has a length of 14.5 km with approx. 400 load pick-up and drop-off stations. There are 74 AGVs in operation, and more than 2000 trolleys in two different designs are used as loading aids. For each AGV, either two small trolleys are used to pick up containers up to DIN size, or one so-called oversized trolley is used to pick up large containers. In addition, there are the sequencing racks with special superstructures (Fig. 3.14).

To transport a trolley, the AGV drives underneath and lifts it up. The vehicles travel at a maximum speed of 1.2 m/s in the main direction of travel. A laser scanner, which monitors the area in front of the vehicle, takes care of personal safety and obstacle detection. The vehicles drive backwards only to position themselves—at a maximum speed of 0.3 m/s and with an acoustic warning signal switched on. A safety edge attached to the rear of the vehicle prevents employees from being injured by the reversing vehicle.

Economic efficiency calculations and simulations were already carried out in the early planning phase. Manually operated vehicles, such as tugger trains, were the main competitors of the AGV. The long distances from the warehouse to the assembly area spoke in favour of automatic vehicles; in addition, for reasons of quality and operational safety, the assembly should be free of forklift trucks. The AGV convinced with the consistency of the concept and the sustainability due to flexibility and the absence of damage to peripheral equipment. In addition, the profitability calculations of the AGV solution showed the fastest return on investment.

The automated guided vehicles find their way with the help of magnet grid navigation. Approximately 3000 permanent magnets are embedded in the ground at regular distances of about 5 m along the entire drive tracks of the AGVs. These cylindrical magnets have a

Fig. 3.15 An AGV takes over a rolling container with audio components. (Source: DS AUTOMOTION)

diameter of 20 mm and a height of 10 mm. They are detected and evaluated by a magnetic sensor bar attached to the underside of the vehicle as it passes over them. The vehicle-side dead reckoning navigation based on (non-driven) measuring wheels is improved in its accuracy by the use of a fibre-optic gyroscope, so that a vehicle's positioning accuracy of ±5 mm is achieved. This is particularly necessary at the load transfer points during automatic load pick-up and delivery (Fig. 3.15) as well as when entering the battery charging stations (Fig. 3.16).

The implementation of free navigation is necessary in this application because the drive track layout is characterised by many overlapping curve radii. Such a layout cannot be realised with inductive track guidance or any other physical track guidance method.

3.2.1.3 Logistics Task at Deutz AG in Cologne-Porz

The renowned engine manufacturer Deutz in Cologne-Porz produces diesel engines not only for trucks but also for mobile machinery, the navy and agriculture sector. In 2009/ 2010, the main components of a very old AGV system were replaced (so-called retrofit) in the engine assembly area (Fig. 3.17).

The renewal included, among other things, the delivery of 43 automatic pallet trucks with laser navigation. They replaced the previously used AGVs with inductive guidance and are now in use in three-shift operation on a maximum of 6 days a week—on a total distance of about 6800 m and at speeds of up to 1.6 m/s (forward as well as backward).

Fig. 3.16 Left: Detail view on a vehicle with two sequencing racks;. right: Battery charging stations for the NiCd batteries. (Source: DS AUTOMOTION)

Fig. 3.17 Three forklift AGVs of the Deutz fleet. (Source: MLR)

The immense challenge in modernising the almost 20-year-old system was the time available: Within only 12 working days at the turn of the year 2009/2010, the new vehicles had to be fitted with components of the previous equipment, and other important system components had to be installed and put into operation. At the same time, the performance of the system was increased and the master control computer functions were optimised.

The converted vehicles are able to handle a wide variety of loads: For example, the vehicles were given four different lifting forks—some with load centring, because they are used to transport very different load units that can weigh up to 1000 kg (transport boxes, mesh boxes, pallets in longitudinal and transverse direction, large and small engines and

Fig. 3.18 Flexibility in load handling and navigation: the forklift AGVs. (Source: MLR)

also special racks). In addition, the lifting forks can be raised up to a height of 3500 mm to serve the racking system (Fig. 3.18).

Another non-routine job was energy management. Although the previous batteries were adopted in the new automatic high-lift transporters, these had to be extended by two cells in order to increase the battery capacity. In terms of design, Deutz broke new ground: the vehicles were manufactured in a pure white to meet the increased demands of car manufacturers in terms of cleanliness in the production area.

The AGV master control system was also completely renewed and had to be integrated into the complex computer world of Deutz. Some important characteristics are: Highest availability through hot standby operation, management of 1600 load transfer points and connection of further numerous peripheral devices, such as conveyor systems, transfer tables, visualisation systems and terminals. In total, the system handles more than 7000 transports per day.

3.2.1.4 Front-End Assembly at BMW AG in Dingolfing

Among other things, front-end assembly for the BMW 5 and 7 series takes place in Dingolfing. There are two AGV drive track layouts for this, each with 20 AGVs. The maximum payload is 400 kg (Fig. 3.19).

Since the layout of the system is simple and should remain unchanged for years to come, inductive power transmission (see also Sect. 2.3.4.2) was chosen. The vehicles thus receive

Fig. 3.19 The assembly vehicle at BMW in Dingolfing. (Source: dpm Daum+Partner Maschinenbau)

Fig. 3.20 20 AGVs in a simple assembly layout. (Source: dpm Daum+Partner Maschinenbau)

the required power contact-free via the power cables embedded in the ground; at the same time, these cables are used for track guidance (Fig. 3.20).

The machine travels through the assembly area at a slow 30 m/min. Stops are made at the individual work stations, and after the work process is completed, the worker restarts the AGV by pressing a foot switch so that it moves on to the next station. At the end of the

Fig. 3.21 30 AGVs in a simple assembly layout for cockpits. (Source: SEW Eurodrive)

assembly process, the finished front end is measured by a robot and transferred to an EMS, which transports it to the main line for installation.

3.2.1.5 Assembly Line for Cockpits at VW in Wolfsburg

In Hall 12 of the VW main plant in Wolfsburg, the production of cockpits for the SUV model Tiguan is running on an assembly line. This line is based on an AGV with 30 AGVs, which are also equipped with inductive energy transfer technology, and has been producing about 450 cockpits per day since March 2008.

The inductive energy transmission for the supply of the AGVs was a firm specification from Volkswagen, because they had already gained good experience with the wear- and maintenance-free technology in previous projects (Fig. 3.21).

The task of the AGV is relatively simple. On a circuit with a track length of 190 m, the assembly trolleys pass through various stations where the cockpit of the Tiguan is assembled piece by piece. At the end of the cockpit assembly, the track runs parallel to a skid system on which the car bodies are conveyed. A robot applies a bead of adhesive to the cockpits, after which the assembly trolley moves to the installation station. Here the cockpits are removed by a handling unit and inserted into the car body. The empty assembly trolley is "loaded" again with the basic module of the cockpit a few metres further on and then the assembly process starts again.

The AGVs consist of a towing vehicle and a parts carrier trailer. All operating equipment, including the cockpit holder, is optimally adapted to the workers, and the production sequence is designed so flexible that the cockpits of several vehicle types can be produced on the same line. As a matter of principle, there is no need for transfer operations, i.e. system changes of the workpiece carriers in the conveyor circuit. In addition, in the event of malfunctions or failures, the individual components—AGV tractor unit and parts carrier trailer—can be easily exchanged within the line. This increases the availability of the entire system.

The vehicles do not travel through the assembly line at a continuous speed, but the line is divided into four speed zones. In flow production, the trolleys have to maintain the corresponding cycle speed. In addition, the foreman has the possibility to flexibly adjust the cycle speed. Between the last assembly station and the maintenance station in front of the gluing robot, the AGVs make a "fast run" of 0.5 m/s.

The trolley then enters the gluing and mounting station again at a low speed of 0.05 m/s. Exact positioning of ±2 mm is necessary at both stations, even if the final positioning is done by a positioning frame. Once the trolley is free of its load, it travels again at maximum speed to the start of the assembly line (Fig. 3.22).

Incidentally, the AGV was not created by one of the classic AGV suppliers, but was realised in a joint project between VW's specialist departments and a system supplier for drive and energy technology. Volkswagen's goal was to generate a standard for similar assembly lines for the future based on the project experience.

3.2.1.6 Use of AGVs in Car Seat Production

Since 2009, Toyota Boshuko has been using an AGV to automatically feed individual components for seat production at its Somain plant in France. The system consists of 11 vehicles that are used in two production lines.

The client's requirement included that all individual components such as foams and sheet metal parts are to be delivered without additional load carriers according to the so-called Minomi principle. Minomi refers to a process in which the produced parts are transported directly, e.g. with a gravity roller conveyor, onto a mobile rolling rack and then transported onwards by the automatic transport vehicle without intermediate storage and multiple handling.

A highly flexible material handling system was necessary to implement the project. The individual components are delivered directly to the respective production lines, the load transfer is done fully automatically and purely mechanical according to the gravity principle by so-called "shooter racks".

As materials such as moulded foam are usually very difficult to move, high-quality roller rails are used to ensure that gravity transfers can function in a production-safe manner.

Fig. 3.22 A combination of tractors and assembly AGVs. (Source: SEW Eurodrive)

In another production part of the plant, underrun AGVs with a tow cylinder device are used. A material trolley pre-commissioned by an employee is driven underneath and "dragged along" by the AGV and finally deposited on the production line (Fig. 3.23).

3.2.1.7 Use of Underrun AGVs for Production Supply at BMW
The project described below is certainly not a typical AGV project in terms of its initial situation and also its objectives, but it should also be presented here because of the development it initiated in the market—one could even speak of a competition among some AGV manufacturers.

The Initial Situation
In the automotive industry, material transport and supply with manually controlled tugger trains are widespread: a tractor with usually four coupled trailers shuttles between the so-called supermarket, where the transport containers are filled by order pickers in precise sequence, and the consumption/assembly locations along the production line. In addition to route finding and in-time delivery (Just-in-Time—JiT, Just-in-Sequence—JiS), the tasks of the tugger train driver also include the exchange of the transport containers standing on roller carts (RC) at the points of consumption. From an ergonomic point of view, pulling and pushing the RUs, which can weigh up to 1000 kg in exceptional cases, is a considerable strain on the tugger train drivers. In case of 4-trailer transports, there is also the challenge of optimising the overall route for all four containers while ensuring JiT/JiS delivery. In a joint R&D project of the BMW Group and Fraunhofer IML, automated transport vehicles were

Fig. 3.23 Underrun AGV tows material trolley with lateral gravity transfer system, so-called "shooter". (Source: CREFORM)

developed in 2016 for the automated individual transport of roller carts and thus as an innovative alternative to tugger train transports.

The Requirements
The following specifications were particularly important for the project:

- As the AGVs were to drive completely under the RC for transport and lift it free from the ground, the max. vehicle height was limited to approx. 22 cm and the max. vehicle width to approx. 60 cm (Fig. 3.24). Due to the very large number of RCs available in the BMW group worldwide, no modifications of the RCs were accepted and thus the mentioned dimensions for the AGVs were set. At the time of the project start, no AGV with this low overall height and a payload of up to 1000 kg was available on the market.
- In order to keep the costs per AGV as low as possible, the strategic decision was already made before the project began that BMW would (co-)develop the vehicles and then manufacture them itself. In addition, as many components as possible should be used in the AGVs that are also installed in the BMW Group's passenger cars (e.g. the lithium battery or camera-based assistance systems).

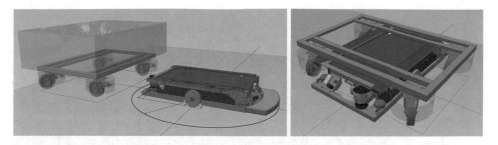

Fig. 3.24 Underride AGV for transporting roller carts. (Source: Fraunhofer IML)

- As little infrastructure as possible should be installed for vehicles' track guidance and navigation, and in particular no installation should have to be made in the ground.
- The pick-up of an RC should be automatic, as should the setting down of an RC at its destination, and the RCs should not be bound to fixed positions by any guide rails or other mechanical aids.
- The AGVs should have a certain degree of autonomy in order to be able to react independently to unplanned obstacles that appear on the roadway in front of the vehicle by taking evasive action or driving around them.

The Solution

A hybrid positioning technology is used, consisting of dead reckoning (measurement of wheel revolutions) and innovative radio localisation (Kinexon, Munich), in which stationary so-called "anchors", mounted on hall walls and pillars, serve as reference transmitters for two receivers ("tags") on each vehicle. In this way, the X and Y position as well as the orientation (heading) of the vehicle in the hall's coordinate system can be determined in real time and with sufficient accuracy while driving. To meet the increased accuracy requirements before and during automatic load changes, a 3D laser sensor system is also used to measure the vehicle position relative to the RC or the gap in which it is to be deposited (Fig. 3.25).

In the meantime, numerous AGVs manufactured and commissioned by BMW employees themselves under the product name STR (Smart Transport Robot) are in use at several locations. To our knowledge, however, it has not yet been finally decided whether BMW will maintain its commitment in this regard on a permanent basis. The special situation that the manufacturer and operator of the vehicles are identical offers, on the one hand, the (rapid) opportunity for continuous change/improvement of the product and its characteristics, but on the other hand also requires dealing with a multitude of details that an "operator only" does not have to worry about:

- Technical development of standard mechanical and electrical components
- Development of technical highlights

Fig. 3.25 Underride AGV for the transport of roller carts; above: prototype (built by Fraunhofer IML), below: series vehicles built by BMW itself in use in Wackersdorf. (Source: BMW)

- Planning and realisation of an overall system (AGV) that is to fulfill performance and availability on a permanent basis
- Commissioning and acceptance, i.e. in particular the execution of performance and availability tests
- Assumption of the role of the manufacturer with the obligation for CE marking, guarantee, warranty, maintenance, servicing

These points are easily underestimated, but are equally important for applications within the BMW Group as well as for sales to customers outside the BMW Group.

The chosen approach is probably out of the question for the majority of AGV operators, but it is tried again and again when there is a foreseeable large number of vehicles to be procured that are not available on the market in the desired form or with the desired range of functions.

The "competition" among AGV manufacturers mentioned at the beginning has led to the fact that there is now a selection of several comparable flat underrun AGV models available on the market.

Fig. 3.26 AGV as a tractor based on converted series devices. (Source: E&K Automation)

3.2.1.8 Improving Production Efficiency at Denso in the Czech Republic

Denso Manufacturing in Liberec in the Czech Republic is a company of the Japanese DENSO Corp, a globally active automotive supplier. Air conditioning systems for passenger cars are manufactured at the site in the Czech Republic. By automating and thus optimising the transports within the production halls using AGVs, the efficiency and flexibility of production could be increased (Fig. 3.26).

After intensive planning analyses, the decision was made in favour of an AGV concept with two Linde P30C tractors with a trailer capacity of max. 3000 kg. It was important to integrate the unmanned tractors as easily as possible into the complex production system. Another requirement was that the tractors could also be operated manually if necessary.

In addition, one of the two existing tractor lines was to be completely saved and downtimes eliminated.

Eight production lines are served by two vehicles and connected to the warehouse; a tugger train can run through up to 20 cycles per shift. The vehicles have a small remote control with which they are precisely positioned (manually) at the loading and unloading point. To guarantee maximum flexibility, each vehicle can serve each of the assembly lines. There is no AGV master control system; despite the existing intersections and junctions, it was possible to do without it completely. The vehicles are assigned manually at the operating terminal at the touch of a button.

The tractors are guided by an inductive wire buried into the floor. The drive track layout is not very complex and forms a circuit, which is why inductive guidance was chosen.

The tractors can also be used outside the closed system: With the push of a button on the terminal, the operator can switch from automatic to manual control. The tractors are thus flexible and can also be used for other tasks.

The amortisation of the system pays off in less than 2 years through the savings in operating staff. Two employees per shift can be deployed differently today—that is six full-time equivalents in three-shift operation.

3.2.2   Paper Production and Processing

Paper roll handling was one of the first AGV applications in Europe. As early as the 1970s, there was a desire to automate the handling and transport of the precious paper rolls from the warehouse to the printing presses. A major reason for these efforts was to avoid damage to the roll, which often rendered the outer paper webs unusable during conventional manual handling—not infrequently up to 10 cm in outer diameter of a roll!

3.2.2.1 Transport and Handling of Paper Rolls at Einsa Print International

Einsa Print International is one of the leading companies in the Spanish printing industry. It produces catalogues, magazines, registers and directories. Since 2007, it has successfully optimised its production and storage by using AGVs.

The system configuration at Einsa—a combination of vertical paper roll storage and horizontal delivery of the paper rolls to the presses—is typical for the paper industry. Thanks to the integration of a so-called downender system on the AGV, it is no longer necessary to install a swivelling device on the conveyor, which results in a considerable saving of space and costs. In addition, the rotation of the rolls during transport shortens the production time (Fig. 3.27).

The Downender system is a technology developed in-house by Dematic that allows the AGV with hydraulic clamps to lift heavy paper rolls vertically, rotate them during transport and deposit them horizontally into the printing machines. Three AGVs with downender function and four fork-lift AGVs run in the plant, transporting printed and cut sheets of paper on pallets to the production machines.

Depending on the size of the rolls, a downender AGV stacks up to four rolls vertically on top of each other (up to a height of 6 m). When the software sends a transport command, one of the vehicles picks up a roll and brings it to the target position in the production line. Here the roll is deposited in a horizontal direction. From this place, the fork-lift vehicles pick them up again and transport them to the presses.

3.2.2.2 Newspaper Printing at the Braunschweig Printing Centre

The Braunschweig printing centre is part of the WAZ Media Group and produces numerous daily newspapers. Four-colour printing is done on three offset rotary presses. In addition to standard newsprint, there is a choice of various higher-quality paper grades.

In 2007, an existing AGV System was replaced by a new one. The commissioning was carried out during ongoing operations. The three AGVs have a lifting frame equipped with two forks that are continuously adjustable in the clear width of up to 700 mm. This makes it possible to transport different types of paper rolls. The largest possible roll has a diameter

Fig. 3.27 Paper roll transport with multifunctional AGV. (Source: Dematic)

of 1500 mm, a length of 1280 mm and a weight of 2000 kg. The rolls can be inserted into and removed from a shelf up to a height of 3.5 m (Fig. 3.28).

The unit is used for the following application: Paper rolls are prepared during the day for printing in the evening. The side parts of the paper rolls are cut off at an unpacking station in order to store them in the paper day storage. By scanning the paper roll, a transport order is transmitted to the AGV master control system. During this time, the transport orders to the reelstands of the printing press are also carried out (Fig. 3.29).

The main production starts shortly before midnight, with the AGV having already filled the reelstands with paper reels from the day's storage beforehand. During production, of course, the reelstands must continue to be supplied with new reels. At the end of a production run, reels that have been started and not used are returned to the day store (Fig. 3.30).

The vehicles are in operation for 16 h every day, after which 8 h are available for automatic battery charging. The vehicles contain 48 V lead-acid batteries with a capacity of 420 Ah. The vehicles have an unladen weight of 3600 kg, the driving speed is 1.2 m/s for forward travel, and reverse travel to approach the pick-up and drop-off positions is carried out at crawl speed (0.3 m/s).

The navigation of the AGVs is done by means of magnets embedded in the ground, the entire layout has a length of 500 m.

Fig. 3.28 A rack storage system as a paper day storage system. (Source: DS AUTOMOTION)

Fig. 3.29 An AGV supplies the staging areas. (Source: DS AUTOMOTION)

Fig. 3.30   An AGV supplies the reel splicer on the press. (Source: DS AUTOMOTION)

3.2.3   Electrical Industry

The electrical industry is representative here for manufacturers of small, high-quality series components. The quality requirements often include extreme cleanliness and order. The weights to be transported are usually not high; standard boxes with a base area of e.g. 600 × 400 mm are often used (small load carriers, bins).

In these productions, flexibility is important: both the layout and the processes change frequently in the sense of continuous process optimisation. IT penetration of the processes is given, a WLAN is usually available and there is less reluctance to use automation technology than in other industries. Small maneuverable vehicles with free navigation are in demand (Fig. 3.31).

3.2.3.1 Just-In-Time Container Transport at Wöhner
Product quality and design are the focus at Wöhner GmbH & Co KG in Rödental. This includes ultra-modern, extremely clean and attractive assembly and flexible intralogistics based on an AGV system. Wöhner is a supplier of innovative busbar systems, load switches, fuse switches and fuse holders for electrical engineering (Fig. 3.32).

Within 4 years, the functionality and the vehicle fleet of the AGV were successively expanded, so that since the beginning of 2010, two vehicle types have taken over the entire production supply. Five vehicles transport the small boxes (600 × 400 mm) and two more

Fig. 3.31 AGV for transport of bins in the electrical industry. (Sourc: left MLR, right DS AUTOMOTION)

Fig. 3.32 Material delivery in the assembly area, AGV with twofold load handling device (left) and for large containers (right). (Source: FROG/Oceaneering)

the large ones (800 × 600 mm). The now total of seven AGVs take over the complete automatic supply of the assembly with individual parts, assemblies and finished products from the container warehouse and the miniload warehouse (AKL).[8]

The main reason for automating the transports was to relieve the employees who used to manually distribute the palletised containers to the workplaces. Today, the delivery of the containers from the existing and the new container warehouse to the approximately 80 workplaces with over 1000 storage locations is carried out exclusively by the AGVS. The vehicles transport about 700 containers per shift.

Due to the passive load transfer and the associated low costs for the transfer stations, the system price could be kept low; the payback period of the initial investment dropped to less than 2 years. The freed-up personnel were relieved of the heavy physical work and put to productive use. Wöhner also took advantage of the opportunity to integrate the AGV

---

[8] AKL = German acronym for automated small parts warehouse.

system as a visual highlight in the attractively designed production environment. The special feature of the vehicles is their load handling. A height-adjustable telescopic belt conveyor was developed especially for container handling, which makes it possible to place containers at different levels on smooth surfaces (shelves, tables, etc.) or on passive roller conveyors and to pick them up again from there.

There is no storage space management for the racks. This is not a problem for the AGVs, however, because a vehicle searches for a free space within the assigned shelf area by means of its own sensors and places the container there. If all the racking spaces are occupied, the vehicle sends a message that is displayed both locally and on the AGV master control system.

3.2.4   Beverage/Food Industry

In this section we want to turn to the beverage and food industry. The price war in these segments is considerable, which is why interest in the use of AGVs has increased significantly in this industry in recent years. What motivation or cost-saving potential lies in this will be explained in the following. To this end, we will first take a closer look at the beverage industry from an AGV perspective before presenting specific systems from the food sector.

3.2.4.1 Intralogistic Optimisation Approaches in the Beverage Industry

The beverage industry is under enormous cost pressure. After intensive investment in production technology over the last 20 years, the remaining major potential for savings is now being discovered in intralogistics. Efficient production facilities must be adequately supplied with empties and auxiliary materials, and the finished products must be transported promptly, reliably and quickly to intermediate storage and distribution.

Historically, the intralogistics of breweries, bottlers and beverage wholesalers have been based on conventional manually operated forklifts. In the past, high transport and storage performance with maximum flexibility were arguments in favour of using forklifts, especially due to the lack of mature alternatives for AGVs.

General Conditions in Beverage Logistics

The high transport capacities are necessary because bottling plants produce ever higher output. For example, a typical beer bottling line today has an output of up to 15 hl/h, which is almost forty pallets per hour that have to be transported away, often around the clock. Since a brewery usually has more than one line in operation, the transport volume increases accordingly. By using suitable attachments, the forklifts are able to transport up to eight pallets at a time.

The full pallets must therefore be transported from the filling lines to the finished goods warehouse at high frequency. Usually, the forklifts bring the pallets to the floor block warehouse, where they are stacked several times on top of each other—without racks.

Stacking heights of up to 10 m are not uncommon. The advantages of block storage are the low system costs and the fact that the means of transport (forklift truck) can enter the warehouse directly, without moving or interrupting, and store them there itself.

Forklifts and block storage have also always been the guarantors of maximum flexibility. In the past supported by human dispatchers, today often by material flow computers, this combination is ideally capable of adapting as quickly as possible to changes and situation-related requirements. We are not only thinking of the block warehouse for the produced goods, but also of the empties warehouse or the warehouse for external products or auxiliary materials. Such warehouses are often even located outdoors and require a high degree of flexibility from the transport system.

This would actually speak for the use of forklifts and the storage concept "block storage"—if it weren't for the limited handling capacities and space resources of block storage and, above all, the well-known disadvantages of forklifts:

- High personnel costs: Especially in the beverage industry there are high pay scales. The annual costs for a forklift truck driver can amount to more than 50,000 € per year. If one wants to operate a forklift around the clock, at least four forklift truck drivers are usually required. If you add the pure forklift costs of at least 10,000 € to the personnel costs, you are dealing with annual costs of more than 200,000 € for a forklift.
- Human error leads to unreliable transports, damages to the forklifts, the product and the surrounding equipment. In addition, accidents, sometimes even involving personal injury, do cost time and money.
- Traceability of the products shipped: These timely requirements—especially in the environment of the food industry—also speak for more automation.
- The high transport performance of forklifts mentioned must also be questioned: There is often a serious difference between the average and the peak performance. Automation technology works much more continuously, reliably and thus more predictably.

Alternative Solutions

The logistic issues are ultimately:

1. New construction of an automatic high bay warehouse (AS/RS[9]) instead of the floor block storage?
2. Replacement of conventional forklifts with AGVs, EMS or stationary materials handling technology?
3. Mixed operation of several systems or consistent, uniform storage and transport systems?

---

[9] AS/RS = Automatic Storage & Retrieval System.

The advantage of an AS/RS lies in its high, reliable system performance. In a comparatively small area, this self-contained system operates automatically, safely and with high availability. However, a system transition is required, no matter how the supply and disposal is technologically solved. Transfer positions must be created, probably additionally with a buffer function. In this context, we want to understand AS/RS as fully automated warehouses, where storage heights of 10 m are exceeded many times over, thus significantly reducing space requirements. Storage and retrieval machines (SRM) are an integral part of the AS/RS.

Manual or automatic storage with forklifts, on the other hand, is feasible at heights of up to 10 m, depending heavily on the load weight.

The location plays a decisive role in the decision between AS/RS and block storage. On the one hand, the site boundary conditions may mean that the construction of an AS/RS is not possible from a structural engineering point of view; on the other hand, the construction of an AS/RS may be unavoidable because the space is not available for upcoming expansions of the block storage facility. In addition, an AS/RS facility is considerably higher than a block storage facility, namely at least 12 to a maximum of 50 m.

The advantages of the block storage system primarily lie in the low system costs and in the fact that both forklift trucks and, alternatively, automated transport vehicles can handle storage and retrieval directly. A change of transport system—as with the AS/RS first to a stationary conveyor system and then to the stacker cranes—is not necessary. This saves time and money. On the other hand, decoupling the transport and storage systems also has considerable advantages. It is realised through dedicated buffer locations on the floor or on special conveyor sections.

On the one hand, the transport system is not burdened with time-consuming storage and retrieval operations, on the other hand, the buffer spaces ensure safety in case of any system malfunctions. Finally, transport systems have their strength in bridging (long) distances quickly and storage systems in overcoming height while covering rather smaller distances. In this way, an AS/RS achieves a far higher number of storage cycles than a transport system of any kind could.

However, if the warehouse turnover is low or the warehouse aisles are too short, an AS/RS is not worthwhile. Slowly rotating warehouses make an automatic high-bay warehouse with stacker cranes uneconomical. The high speed of the SRM does not come into play and the required number of SRMs makes the AS/RS too expensive. In such cases, the seamless solution (block storage) has advantages again.

For the transports to link the production with the warehouse and the incoming and outgoing goods, not only forklifts and AGVs come into question, but also the electric monorail system (EMS) as well as stationary conveyor technology, such as roller conveyors or chain conveyors. Figure 3.33 shows the sources and sinks of the material flow.

The illustration also shows any load transfer stations that may be required. These mean the provision or intermediate buffering of goods. A system change is thus unavoidable both

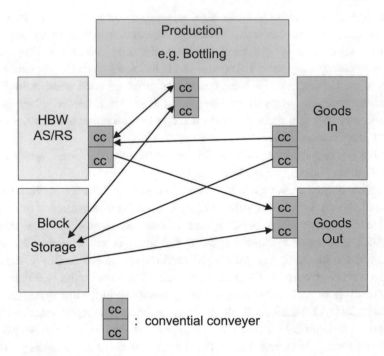

Fig. 3.33 Typical flows of goods in a brewery. (Source: Ott and Ullrich)

after production and before the high bay warehouse. Only the block storage system can be operated directly by forklift trucks or AGVs.

On the right side of the picture, incoming and outgoing goods are sketched, also with load transfer stations. These seem advisable in any case if there is automatic internal transport. Because then, with these load transfer stations, the strict separation of automated transport and truck loading and unloading by means of forklifts succeeds. Although there are already first examples of automatic truck loading and unloading with AGVs, this is still far from being standard.

In the future, however, even these types of AGVs will not be the same as those used for mass transport. The specific requirements are too different here, so that in each case a transfer from one technology to the other must take place. If one also takes into account that the incoming and outgoing goods do not take place all day compared to production and that the loads have to be made available to the trucks in shipping zones, the transport routes are broken up. It cannot be assumed that loading will take place directly from the warehouse onto or into the truck in near future. Therefore, shipping zones, gravity roller conveyors or other stationary conveyor technologies will futher on interrupt the transport.

From today's point of view, the following division is recommended: the automated world inside the brewery and the manual forklift area outside with the trucks and the outdoor storage areas (e.g. for empties). This is because automation is not as obvious for outdoor use as it is indoors.

Table 3.2 Technical suitability of materials handling solutions according to VDI 2710, Sheet 1

Criterion	Forklift	AGVS	Conv.	EMS
Task flexibility	++	+	−	O
Layout flexibility	++	++	−	−
Nominal performance	+	+	++	+
Peak performance	++	O	+	O
Concealed space	++	++	−−	−
Ceiling load	++	++	+	−−
Personal security	−	++	+	+
Order and reliability	−	++	++	++
Round-the-clock operation	−	++	++	++
School grade	**2.1**	**1.4**	**2.4**	**2.8**

++ very good; + good; O satisfactory; − sufficient; −− deficient

In principle, the following techniques are conceivable indoors: Stacker versus AGV versus EMS versus continuous conveyor technology. The system decision has technical and economic aspects. The VDI Guideline VDI 2710 Part 1 "Decision Criteria For The Selection of a Transportation System" helps with the technical system selection. The evaluation table (Table 3.2) is based on this guideline.

The table gives a first impression of the impact of the relevant criteria. However, a site-specific adaptation of the table is necessary. In general, there is nothing to prevent the use of several systems in combination, which can certainly be applied when expanding existing facilities. For example, an existing block storage structure can continue to be served with manually operated forklift trucks, while the removal from filling lines to the warehouse is carried out with AGVs. In this way, continuous, low-disturbance disposal of the filling lines can be achieved, while the block storage operation, which is more prone to disturbances, is decoupled from filling, e.g. by stationary conveyor technology as buffer sections.

At this point it should be mentioned that mixed operation of manually operated forklift truckss and AGVs in the same layout is also possible. Clear rules are required here (e.g. "the AGVs always have priority") and the forklift truck drivers must be intensively prepared for their new automatic colleagues. Nevertheless, the forklift-free factory is the more consistent and safer logistics solution.

Suitable AGV Concepts

Basically, two different types of vehicles can be used: the piggyback vehicle (Fig. 3.34, left) and the fork vehicle (Fig. 3.34, right). Piggyback vehicles are equipped with conveyor elements (roller conveyor or chain conveyor) and take care of the pallet pick-up/transfer laterally to stationary conveyor equipment. The typical features of both types of vehicles are summarised in Table 3.3.

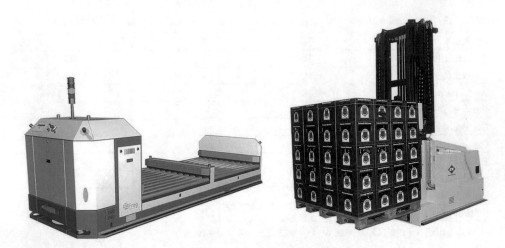

Fig. 3.34 AGV types for the beverage industry: on the left a piggyback AGV, on the right a fork truck. (Sources: left FROG/Oceaneering; right E&K Automation)

Table 3.3 Feature comparison of the two different vehicle types

Technical feature	Piggyback AGV	Fork AGV
Floor-level pallet pick-up	Not possible	Possible
Stacking pallets	Not possible	Possible
Space and time Requirements for load handling	Low, because positioning in direction of travel with lateral load transfer possible	High, because positioning and load transfer is required at right angles to the main direction of travel
Driving speed	Max. speed on long straights up to approx. 2 m/s	Same as Piggyback AGV
Kinematic principles	All variants can be used	Vintage 3-wheeler, possibly with three steered wheels
Navigation technology	Common systems can be used	Restrictions in block storage that can be overcome by special solutions; new, innovative processes are in demand
Space requirement in the layout	Hardly bigger than the load	Significantly larger than the load, especially with the counterweight version
Manual operation	Hardly suitable	Conditionally suitable, depends on the design
Flexibility of use	High	Very high, as no stationary conveyor technology is required

With the aim of standardisation, one often relies on a single vehicle type per operational area of a location. When choosing one of these, the importance of the block storage is one of the most important criteria. If a block storage system is or has to be used, it should also be served directly by the AGVs, which answers the question of the vehicle type.

If it makes sense to install conveyor technology, e.g. to decouple different systems or to separate transportation (e.g. AGV) and storage technology (e.g. SRM), the choice of piggyback vehicles provides more consistent automation. In addition, a buffer function is achieved with somewhat larger load transfer points (longer conveyor routes), which brings more safety to the overall process (Table 3.3).

Summary

There is no such thing as the optimal logistics solution in the beverage industry; the conditions at the various locations are too different for that. Often you will find grown structures in historic plants in the middle of a city. Only rarely one will have the luxury of being able to plan anew on a greenfield site. The building blocks of modern intralogistics are listed above. The optimal solution can be put together for each individual case.

3.2.4.2 Project Example Radeberger Brewery

At the Radeberger brewery in Dortmund, beer is brewed around the clock, which means that in 24/7 operation, pallets loaded with crates of beer have to be taken from production by roller conveyors and transported to a storage area of approx. 9000 m². In the warehouse, up to four pallets are then stacked on top of each other within storage blocks corresponding to different products and production batches. In this way, a total of more than 10,000 pallets can be buffered and made available for shipping/truck loading.

The transports of two pallets at a time from the roller conveyors into the block storage areas, including the stacking processes, are carried out by a total of 17 AGVs; the subsequent truck loading continues to be carried out in the traditional manner by manually operated front stackers.

A camera and image processing system specially developed for this purpose is used to precisely stack the pallets on each other. In combination with laterally movable forks, the required accuracy can be reliably achieved.

The omni-directional vehicles are based on the counterweight principle, i.e. with freely cantilevering forks, determine their current position inside the storage blocks by means of magnetic grid navigation (see also Sect. 2.1.1.2) and outside the blocks by means of contour navigation (Sect. 2.1.1.4). The entire use of contour navigation and thus the complete elimination of any ground installation is not possible because the "environment" inside the storage blocks is too volatile, i.e. subject to too strong and too frequent changes (Fig. 3.35).

3.2.4.3 Innovative Order Picking at Marktkauf Logistik GmbH

Marktkauf Logistik GmbH, headquartered in Bielefeld, unites all warehouses for the optimal handling of goods for the EDEKA Group. At the Laichingen location, a concept

Fig. 3.35 AGVs in operation at the Radeberg brewery. (Source: E&K Automation)

called "Logistics-by-Voice" in combination with an AGV system was used for order picking in the warehouse until 2008. At the time of the system's introduction (2003), the solution described below was an absolute innovation and the change to a different working principle after only 5 years of operation had no technical but company policy reasons. The system can be seen as a model for several other implementations and offers from other suppliers to support order picking through the use of AGVs—and that is why it is also presented here.

No other market is as famously competitive as that for groceries: margins are tight, low costs for logistics play a key role. This led Marktkauf to the decision to effectively modernise its logistics: All warehouses combined in the company are gradually receiving solutions shaped by new technologies for the economically improved handling of goods (Fig. 3.36).

The Marktkauf warehouse in Laichingen started with a model solution. The top priority was the analysis of individual functions, especially for the planning of an efficient system solution in order picking. Various surveys as well as calculations over a longer period of time had the result that many ineffective manual operations were associated with the way of order picking as it was done before: The order lines had to be crossed off long lists item by item. Even the order lines that could be read from screens required confirmation by hand. In addition, the long distance rides of the picker to the goods-out area were a major disadvantage.

This conventional way of picking was thus characterised by time-consuming secondary activities. Not only the close observation of the processes led to this result, but even more

Fig. 3.36 Order Picking AGV, optionally also for manual operation. (Source: E&K Automation)

the calculated numerical values. According to these figures, the order pickers spent only 3 h in a 7.5-h shift on their actual activity of compiling collis[10] from the shelves into finished shipping units.

With the new concept, the order picker receives his orders in spoken form via radio on his headset. He now has both hands free for his work, especially since his assigned automated transport vehicle already drives ahead to the next order point after the spoken O.K. at the end of a pick. This eliminates the need for the order picker to get on and off the truck and to park the vehicle depending on the position. Even the distances between the vehicle and the shelf have become shorter because the pallets for loading are now always positioned exactly at the shelf. The automatic pick vehicles also take over the automatic trips to a transfer point. Then, after picking up an empty pallet from the store, they immediately reenter one of the predefined racking aisles. With this technology, the picking performance of the order pickers increased by almost 100%.

Already during the registration at the beginning of the shift, an AGV starts moving. Each order then starts with a short announcement by the picker. The voice manager then calls out the number of the first shelf and the pick location for the pick. The system also states the number of packages to be picked. Once the process is completed, the picker

---

[10]Colli (plural) are the smallest units, i.e. usually individual pieces of a consignment of goods.

confirms by speaking "O.k." to the system via his headset microphone and then receives the next order.

When a pick-to pallet is full, it is temporarily secured with wrapping foil. By voice command, the picker then sends the vehicle with the pallet to the transfer point. During this time, a second AGV moves into position for picking the next order. The pick vehicles drive the full pallets to the aisle exit and place them there for pick-up by another forklift AGV. The pick vehicle then retrieves an empty pallet from storage and drives to the designated racking aisle.

The further path of the full pallet leads from the transfer point to the stretcher for film wrapping suitable for shipping. An integrated printer attaches the data sent by the system for the destination to the load in the form of a pallet label. In this way, the finished pallet receives all the important information for shipping, i.e. for staging and truck loading by the forwarder.

In summary, performance has been doubled and the error rate reduced by 60%. In addition, there was a significant improvement in handling ergonomics, as the pick-to pallet can always be adjusted to the optimum loading height in a lifting area. The increase in picking performance is achieved through the combination with an AGV, ultimately through the automation of the collection runs.

As already mentioned at the beginning, there are now other suppliers for this type of picking support—the "iGO neo" solution from Still can be mentioned here as an example (Fig. 3.37). Here, too, manual operation is possible at any time if required, since the basic device is a so-called horizontal order picker with a driver platform and all the operating elements required for manual operation have been retained in the automation.

And yet another solution approach should be briefly presented at this point: If the order picking principle is reversed, i.e. if the goods are brought to the order picker, support by AGVs is also possible. This was first demonstrated in 2009 by the American company Kiva Systems (see also Sect. 2.3.1.5): The AGVs transport shelves with goods (= the articles to be picked) to the order picker's workstation and back to the warehouse (= staging area) after the manual removal of the articles (Fig. 3.38).

Kiva Systems was acquired by Amazon in 2012, has since been called Amazon Robotics, and to date (June 2019) has built and deployed over 100,000 underrun AGVs (they are called "robots" by Amazon) in Amazon's distribution centres.

3.2.4.4 AGV Monitors Cheese Ripening Process at Campina

Cheese manufacturer Campina is a brand of Royal Friesland Campina and operates an AGV system with four laser-guided AGVs in Bleskengraaf/NL. The automatic vehicles transport the cheese wheels within the cheese production plant. In doing so, they move the cheese stacked on racks completely independently between the warehouse where the cheese is ripening and the two processing machines (Fig. 3.39).

The AGV master control system takes over the integrated recipe and warehouse management in addition to the usual functions. This manages the cheese recipes within the process. Each recipe contains a number of specified treatments that must be carried out

Fig. 3.37 Support of the order picker by AGVs. (Source: Still)

Fig. 3.38 Support of the order picker by AGV. (Source: Amazon)

regularly with the cheese. Depending on these defined recipes, the AGVs automatically bring the cheese to the processing machines.

With an advanced warehouse management module, the entire warehouse and every stored good in it is visualised. Depending on the current status of the warehouse position (row occupancy, number of free positions) and the recipe details of the goods, the software determines which load in the warehouse must be picked up first by an AGV.

The AGVs transport cheese pallets with a length and height of more than 2 m each and a width of only 85 cm. These are therefore narrow and very unstable loads, which required the development of a special customised AGV.

The cheese pallets are stored one by one in the rows of warehouses in low storage rows. Ventilation pipes are located to the left and right of the racks. The free space between each cheese pallet and the ventilation pipes is only 5 cm on each side. This means that the AGV

Fig. 3.39 The "Cheese AGV" entering the narrow storage aisle (left) and two AGVs handling the ripening racks (right). (Source: Dematic)

has to ensure very high stability to avoid vibrations and, of course, collisions during entry and exit.

3.2.4.5 AGV Made of Stainless Steel in the Schönegger Cheese Dairy, Steingaden

In order to meet the strict hygiene regulations in food production, vehicles made entirely of stainless steel are particularly suitable. The bright metallic surface repels bacteria and can be cleaned quickly and easily. If, in addition, all panels as well as the control and drive modules are sealed, the vehicles can be disinfected from all sides—even from below—with superheated steam.

The automated transport system installed at the Bavarian cheese producer Schönegger works through a detailed cheese care programme for each of the 120,000 cheese wheels, in addition to the actual transport tasks (Fig. 3.40).

Made of stainless steel, the free-running pallet trucks serve the refrigerated and ripening warehouses, load and unload the cheese care machine and take the racks to dispatch for packaging. When the driverless transport vehicles pick up a stack of cheese wheels, they automatically identify it with the barcode reader and transmit the data to the warehouse management software, which organises the cheese in the ripening warehouse according to batch and ensures that the cheese care programme is processed accurately. This makes each batch traceable along the logistic chain. It can be determined at any time when, where and by whom the goods were received, produced, processed, stored and transported.

Fig. 3.40 The stainless steel pallet truck carries 4.6 tonnes and reaches a lifting height of 3.8 m. (Source: MLR)

3.2.5   Building Materials

Neither construction nor the production of building materials are typical AGV target industries. Nevertheless, there are also application possibilities here, one of which we would like to present here; the production of Styropor insulation boards. This rigid foam is the all-rounder among building insulation materials, used in roofs, walls and floors. As an intermediate product, 5 m monoliths have to be transported. The technical requirements for manoeuvring and handling are high (Fig. 3.41).

The monoliths are created in the so-called block moulds by thermal expansion. There, the huge cuboids (5100 × 1050 × 1300 mm) have to be picked up standing upright and brought to a warehouse. The AGVs are equipped with impressive grippers with which they can pick up one or even two of the blocks weighing up to 230 kg at a time. Load pick-up and delivery can be done variably at ground level or from/on powered roller conveyors up to 500 mm high (Fig. 3.42).

The vehicles transport the monoliths to the block storage area, where they are parked at ground level for maturing. Depending on the end product, such a maturing process takes from 1 day to several weeks. The block storage is operated exclusively by the AGV system. Manual intervention, e.g, with forklift trucks, is not desired because no forklift truck drives as consistently reliable and accurate as the automatic vehicles. The block storage has a high storage density, the distances between the blocks are very small.

After maturing, the blocks are transported to one of the three cutting machines, which is also done by the AGV. There, the blocks are cut so that slabs with the desired final

Fig. 3.41 The AGV with its large grippers for the sensitive giants. (Source: DS AUTOMOTION)

dimensions are produced. The slabs are stacked ready for packaging and transported to the finished goods warehouse on pallets with manual forklift trucks.

The AGV navigates with the help of the magnet point sequence, which makes it possible to master the narrow and at the same time very high block storage. The AGV is a clamp stacker, in which case a customised clamp design is combined with a standardised lifting frame. The clamp has to dose the holding forces very sensitively in order not to damage the sensitive polystyrene monoliths. The lifting frame is able to place the load at different heights (Fig. 3.43).

The vehicles get their energy from lead-acid batteries that last for two shifts. Then the vehicles automatically drive to the charging stations, where the empty batteries are manually exchanged for fully charged ones. For this purpose, the employees have special trolleys at their disposal on which the batteries to be charged are placed.

Fig. 3.42 Load pick-up from the roller conveyor. (Source: DS AUTOMOTION)

3.2.6 Steel Industry

Automated transport systems are not only used in classic areas of intralogistics, they are also increasingly conquering areas where things are a little rougher and more robust.

This also includes the steel industry, from production to processing and finishing. Outokumpu GmbH is one of the world's leading manufacturers and processors of stainless steel. The group headquarters are located in Espoo, Finland, and Outokumpu operates a distribution centre for Europe in NL-Terneuzen. Approximately half a million tonnes of stainless steel are processed and delivered there every year.

The main focus is on a modern, consistent and reliable logistics system. This consists of an automatic warehouse for steel coils, three automated transport vehicles, four production lines, an automatic intermediate storage facility for sheet metal packages and a modern production planning system.

The delivery centre in Terneuzen receives most of the steel in the form of coils delivered by ship. There they are uncoiled and cut to length according to the customer's wishes, and delivered again on time as coils or as bundles of sheets. There are four production lines for this. Two in which the material is cut to the right length and two in which it is cut to the right width.

Automated transport vehicles play a key role in these processes. They ensure the delivery of raw goods from the warehouse to production as well as the return of finished products to intermediate storage or directly to shipping (Fig. 3.44).

Two 30-tonne heavy-duty AGVs, which are designed as mandrel stackers, take over the supply of production from the automatic coil warehouse. The coils are picked up at the

Fig. 3.43 Tight setting situation in the block storage area, a task for the AGV. (Source: DS AUTOMOTION)

Fig. 3.44 The AGV carries a 30-tonne coil directly with pallet. (Source: FROG/Oceaneering)

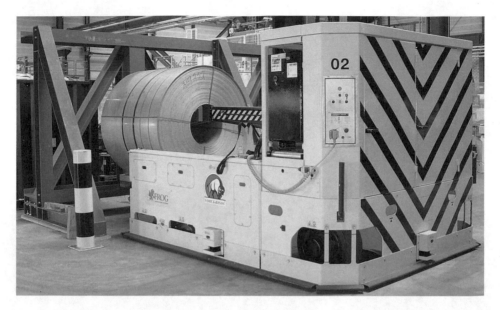

Fig. 3.45 Coil transfer from the automatic mandrel stacker to the turnstile. (Source: FROG/
Oceaneering)

transfer points at the warehouse exit with a so-called mandrel in the centre of the coil.
Hanging on this mandrel, the coil is brought to the corresponding production line and
suspended on a turnstile. The turnstile is operated from one side by the AGV and from the
other side by the production. The turnstile thus serves as a buffer and as an interface
between the mandrel stacker AGVs and production. Coils that have been started with
residual quantities are brought back to the coil warehouse by the two mandrel stacker
AGVs and automatically stored (Fig. 3.45).

The two mandrel stacker AGVs and another 6-tonne heavy-duty AGV, which has a
chain conveyor as a load handling device (Fig. 3.46), take care of the disposal of the
production. Cut-to-size sheets, which are wound into coils and sent to the customer, are
packed on disposable pallets at the respective exit of the production lines and brought to
dispatch by the heavy-duty AGVs. Sheets that go to the customer stacked as a package are
provided on disposable pallets at the automatic transfer stations of the production lines.
There they are picked up by the chain conveyor AGV and transported either to the
intermediate storage or directly to packaging and on to shipping.

The entire production processes are controlled via the production planning system (PPS)
and the required transports are transmitted to the AGV master control system in the form of
transport orders. This combines and optimises the transport orders and assigns them to the
appropriate vehicles as transport orders. All completed transports are reported back to the
PPS. This ensures complete tracking of the goods.

The vehicles have suitable wheel configurations according to their payload and application requirements. The mandrel stacker AGVs are omni-directional, which means they can
travel in all directions without restriction and rotate around any virtual points.

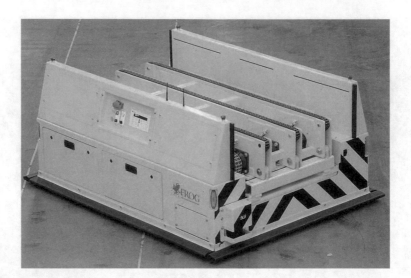

Fig. 3.46 AGV with chain conveyor for pallets with steel sheets weighing up to 6 tons. (Source: FROG)

The vehicles navigate via a magnet point grid laid in the ground. The vehicles plan their own route using a map of the work area stored in the vehicle control system. This map contains the drive tracks as well as the positions, dimensions and functions of all elements that are relevant to the topography of the plant area. These include, for example, walls, doors, gates, lifts, chargers, active and passive transfer stations and all important fixtures and fittings. The map that holds all the elements can be changed by the operator himself at any time.

3.2.7   Hospital Logistics

In the past, little attention was often paid to the logistical processes within a hospital and the associated potential for optimisation and cost savings. However, the financial pressure on hospitals is becoming greater and greater, so that more emphasis is being placed on the overall result. In this context, logistics connects the individual areas—not only technically, but also organisationally and thus ultimately also economically. If one understands how to design these links correctly, undreamt-of economic potentials arise. The logistics manager of a (large) hospital therefore has the task of dissolving parts of the departmental thinking and using it for an overriding overall goal. One must become aware of this fact at all decision-making levels.

AWT[11] systems have always been used in large hospitals. Initially, P&F[12] systems were used, later followed by electric monorail systems (EMS).[13] P&F systems are mechanical chain systems that transport the roller containers[14] beneath the ceiling of the supply aisles. EMS are individual, electronically controlled and electrically driven single hangers, each of which transports a container along a rail under the ceiling.

In the meantime, P&F technology is no longer relevant and has been completely replaced by EMS. Around the beginning of the new millennium, many hospitals worldwide are converting their AWT facilities to AGVs. AGV deals with the main flows of goods in hospital logistics, all of which are transported in roller containers, such as food, laundry, sterilisation goods, pharmacy goods and medicines, magazine goods and waste. Various designs of these roller containers can be seen in the following pictures; the technical requirements have already been described in the second chapter of the book. In large hospitals, a pneumatic tube system is sometimes also available for sending files, samples and other small items in-house.

The advantages of the AGV system over the EMS are:

- Easy installation (during supply operation)
- No ceiling suspensions
- Shared use of existing paths and facilities
- Flexible use, simple reprogramming
- Permanent access to each individual AGV.

In principle, there are these arguments in favour of using AGVs in hospitals:

- Optimisation of logistics processes
- Organised material flow
- Reliable and timely deliveries in the sense of an HACCP[15] concept
- Automatic tracking of material
- Reduction of logistics costs
- Increasing safety
- No damage to containers, doors, walls or facilities
- Integration into existing buildings without interrupting the supply.

Initially, only the very large university hospitals used AGV systems because the economic advantages were obvious. Today, hospitals with more than 600 beds are already considering the use of AGVS.

---

[11] AWT = German acronym for automated transport system for goods.

[12] P&F = Power & Free.

[13] EMS = electric monorail system.

[14] The roller containers are also called transport trolleys or transport containers.

[15] HACCP = Hazard Analysis and Critical Control Points: a preventive system designed to ensure food and consumer safety (1998).

The commonly used vehicle types differ with regard to the load pick-up. Basically, there are the following variants:

- The AGV is basically a fork-type AGV, it has a fork-like lifting device at its rear with which it can drive under the roller container from its narrow side and lift it a few centimetres off the ground.
- The AGV looks similar to a fork AGV, but the roller container is gripped and lifted by the AGV at the top; the transport is also carried out here in longitudinal direction (Fig. 3.47)
- The AGV is so narrow and flat that it can drive completely under the container from its narrow side and only protrudes about 20 cm at its front and rear; for the actual transport there are then again two variants:
  - the roller container is lifted a few centimetres off the ground, i.e. the AGV carries the entire weight
  - the container remains with its wheels on the ground, the AGV tows the container by means of a mandrel that can be pushed upwards (Fig. 3.48)

Forklift AGVs are only used where an AGV has to be linked to an EMS or a P&F and no other solution can be seen.

The vehicle type most frequently used in modern plants and hospitals, on the other hand, is the underrun AGV with lifting device. The advantages of this type of vehicle compared to the forklift are:

- Reduced space requirement: the AGV is only slightly larger (longer) than the roller container, i.e. the container essentially determines the space requirement.
- High manoeuvrability when manoeuvring: the kinematic principle of steering by differential speed with two centrally arranged drives enables tight curve radii and turning-on-the-spot with minimal envelope.
- Fast transfer from automatic transport to manual movement of the roller container.

Underrun AGVs with lift (Fig. 3.49) have prevailed over towing AGVs (Fig. 3.48)—despite their somewhat more complex, because more stable construction—because the precision of the driving manoeuvres and the achievable positioning accuracy with this principle is independent of the quality of the swivel castors of the roller containers and remains permanent.

In the following, two AGV systems in hospitals are presented that were realised according to the principle of underrun AGVs with lifting device.

3.2.7.1 AGV System at Klagenfurt Regional Hospital, Austria

In the Klagenfurt regional hospital (LKH Klagenfurt, 1400 beds), one of the most modern AWT systems to date has been running as an AGV system since 2009. The installation with a large fleet of automated vehicles shows the potential of the AGV as an intralogistics tool.

A new building measure was the reason for a reorganisation of the hospital logistics. The overall project, called "LKH New", has a total duration of 10 years and includes new

Fig. 3.47 A forklift AGV for picking up a roller container from the front. (Source: MLR)

Fig. 3.48 Underrun AGV with mandrel pick-up for towing roller containers. (Source: DS AUTOMOTION)

ward blocks, a central kitchen, a laundry, various other functional areas and a supply and disposal centre. All areas are connected with each other underground, so that a 14 km long network of drive tracks was created. The entire concept is characterised by the fact that great attention was paid to logistics (Fig. 3.50).

Sixty automated transport vehicles take care of the transport of food, laundry, pharmacy and magazine goods. They carry out their transport tasks confidently and radiate a calm, unagitated atmosphere. The reasons for this lie on the one hand in the drive and steering technology, but not least in the navigation method used, magnet point grid navigation.

Fig. 3.49 Underrun AGVs for lifting loads from three different AGV manufacturers. (Source from left: DS AUTOMOTION, MLR and Swisslog, all 2009)

Fig. 3.50 An AGV carries a roller container. The vehicles have a maximum payload of 500 kg. (Source: DS AUTOMOTION)

Safety laser scanners ensure the safety of people and protection of equipment: each vehicle has a scanner at the front and rear that detect obstacles on the road ways even if they are still in a distance of some meters. The vehicles are thus able to adapt their driving speed to the conditions and to integrate themselves into the operation. These safety devices, which are approved by the German Social Accident Insurance Institution, reliably prevent any collisions with people and equipment.

The AGVs are completely made of stainless steel. All panels and the lifting device are sealed on all sides. This means that the vehicle can be easily cleaned and disinfected from all sides—even from below—and thus meets the strict hygiene regulations.

Fig. 3.51 Waiting for the lift. The loading and unloading position directly at the lift for dispatching roller containers. (Source: DS AUTOMOTION)

The goods to be transported—whether food, laundry, waste or medicines—are carried in roller containers that can be pushed by the hospital staff and are positioned on predefined touchdown points for automatic transport. These points are marked with plates on the floor and equipped with sensors to detect the roller containers. The staff member then only enters the destination of the transport at an input terminal and the rest will be executed automatically.

The AGV master control system instructs an AGV that is in the vicinity to carry out this transport. The vehicle drives under the container, lifts it up a few centimetres from the ground and drives it to its destination. As it drives underneath, it reads a transponder on the container floor and checks the plausibility of the transport. This prevents incorrect or unauthorised transports, for example a rubbish container being taken to the kitchen (Fig. 3.51; Table 3.4).

Planning Premises
For the overall system to function at all, comprehensive planning was necessary. Two essential premises were the ring concept and redundancies.

Each functional area was designed as a logistical ring. All material flows consistently in one direction: the material arrives on one side, then passes through the functional area and leaves it on the other side. In this way, opposing material flows are avoided and the kitchen and laundry, for example, appear extremely tidy—a prerequisite for high productivity (Fig. 3.52).

Practical planning also includes the provision of redundancies. When designing the processes, the possible failure of all components and resources involved is taken into

Table 3.4 General description of the underrun AGV with lifting device

Dimensions and weights	L × W × H: approx. 1800 × 600 × 330 mm Max. load capacity: 500 kg
Driving speed	1.6 m/s bidirectional, i.e. the vehicles can move in both directions without restrictions
Positioning accuracy	±10 mm
Incline/dexline	Short sections are rideable at 7%, with "gentle" transitions (25 m radius; there is such a slope at the transition of the new building into the existing building)
Navigation	Magnetic navigation
Safety	Flashing light, emergency stop button (front and rear), audible warning device, programmable voice output
Personal safety	Safety laser scanner with approval by the German Social Accident Insurance Institution for personal safety, front and rear with several warning and protective fields
Data transmission	Each vehicle is equipped with a WiFi client
Vehicle kinematics	Steering by peed difference: 2 fixed mounted traction drives +2 front/rear rotatable support wheels (castor wheels) Motors: maintenance-free three-phase drives, brushless AC wheel hub drives, 24 V Wheels: non-chalking Vulkollan tyres
Cover	All exterior covers in stainless steel, from above protection class IP54
Lifting device	– Electromechanically operated lifting platform with load detection interrogation – Stroke: 80 mm – With RFID transponder reader and light sensors for container identification and localisation
Energy concept	The vehicles are equipped with a traction battery (lead-gel, 200 Ah) that remains in the AGV. The battery is charged automatically at battery charging stations using charging contacts on the underside of the vehicle. An automatic battery charging station is available for each AGV

account: How can hospital operations be maintained if, for example, a lift or a conveyor system fails? For each emergency scenario, a plan B must be thought out in advance so that the failure of a technical unit does not lead to unforeseen serious problems later in daily operations.

A special emergency scenario is the fire alarm. If this is triggered, the AGV system also switches to a special mode that ensures that the AGVs behave in accordance with the situation. This includes that the roadways are cleared and the lifts are no longer used. Automatic doors are no longer passed through so that they can close properly.

Example: Kitchen

Additional ward blocks and the new construction of the kitchen necessitated fundamental changes in the preparation and delivery of meals for the patients. The previous "cook &

Fig. 3.52 The underground world of the AGV system. This is where the long transport routes, buffer locations and battery charging stations are located. (Source: DS AUTOMOTION)

serve" procedure[16] was replaced by the modern "cook & chill".[17] "Cook & serve" means preparing the food and then immediately distributing and serving it, which would be impossible with long transport routes because the legal temperature requirements could not be met. When the food reaches the patient, the hot food would have cooled down too much and the cold food would have become too warm.

That's why today they rely on "cook & chill": the food is cooled immediately after preparation in the kitchen according to the HACCP specifications. Both the trays with the food and the roller containers are precooled to 4 °C before they are sent to the wards by AGV. There is an extra cold room for the roller containers. Forty trays fit in one roller container.

So an AGV drives the refrigerated roller container, which weighs up to 350 kg, to the ward and places it in front of a regeneration station. There, the doors of the roller container are opened and it is docked to the regeneration station. Here, the hot food is then heated while the cold food continues to be cooled. After the regeneration process is completed, the food trays are handed out to the patients by the ward staff.

After the meal, the roller containers are loaded with the used dishes and sent back to the kitchen, where they are emptied by the kitchen staff. After each transport, each roller container passes through the automatic container washing facility, and the transport there is of course also handled by the AGV. In order not to have to use more AGVs than absolutely necessary, almost only food transports are carried out during lunchtime; other transports, such as for waste or pharmacy goods, do not take place during this "rush hour". The same applies to the times immediately before breakfast and dinner.

---

[16] Cook & Serve = The meals are served directly after cooking.

[17] Cook & Chill = Meals are chilled after cooking and only reheated and served later—if required.

Fig. 3.53 An AGV leaving a lift; on the left with roller container. (Source: Swisslog)

3.2.7.2 AGV System at St. Olav's Hospital, Trondheim

The new St. Olav's University Hospital in Trondheim, Norway, is a 950-bed facility. Here, outpatient treatment, research and specialist training are integrated functions. To meet the complex transport requirements of the hospital, an AGV system is used for the automatic transport of food, waste, medication, magazines and sterile goods. The underrun AGVs use a drive track layout with a total length of 4500 m.

Most goods are delivered in roller containers from an external warehouse to the hospital's truck loading ramp, where they are ready for transport by AGVs. Storing the containers at the point of use improves control over the flow of goods and reduces inventory (Figs. 3.53 and 3.54).

The loading and unloading operations are carried out automatically by the system and the transport orders are transmitted by means of transponders. These chips enable easy handling by being placed in a special pocket on the side of the containers.

The navigation system used allows any desired change in the movement of the vehicle and the drive track by simple software adjustment; no modifications in the building are required for this. Here, the so-called contour navigation (see also Sect. 2.1.1.4) is used, which does not require any additional navigation sensors, but uses the already existing data of the safety laser scanner for navigation purposes. The method does not require any fixed, artificial markers such as magnets or reflectors.

The automatic vehicles call lifts and open and close doors via a radio network. Sun shading devices in the building are automatically controlled by sensors and provide a needs-based setting, which is required in particular for the navigation technology described. The lighting switches off automatically if no movement is detected. In the

Fig. 3.54 A loaded and an unloaded vehicle meet each other. (Source: Swisslog)

hospital's supply centre, technicians can follow all processes on the screen. In this way, the AGV is part of a "digital" hospital.

3.2.8 Pharmaceutical Industry

An international pharmaceutical company based in the south of Germany has been operating an AGV system since 2005. It consists of 20 pallet trucks with magnetic grid navigation and an AGV master control system.

The system takes over the delivery of raw materials, consumables and supplies from the existing automatic high bay warehouse (AS/RS) to three production areas. At the same time, finished goods are transported from the production areas back to the high bay warehouse. Euro pallets with a maximum weight of 600 kg and a maximum height of 2 m are transported (Fig. 3.55).

In two steps, the facility was expanded in 2010 and 2012 to a total of 20 vehicles, another production area and a second high bay warehouse housed in an adjacent building. Since direct stock transfers now also take place between the old and the new high bay warehouse, the hierarchies of the individual systems had to be rethought.

The company attaches great importance to the fact that the two structurally separate warehouses and their warehouse management systems can be operated independently of each other. For this purpose, the AGV supplier developed a specific solution with which only the two warehouse management systems can be combined with the company's own AGV master control system.

Fig. 3.55 Vehicle no. 1 at the car park and no. 10 at the handover to the warehouse. (Source: FROG/Oceaneering)

With this solution, all customer requirements could be implemented to the fullest satisfaction. The extensions to the AGV were carried out during ongoing operations. First, the new areas were equipped with the required infrastructure and the new warehouse and the Production Planning and Control System (PPS) were integrated. The new, identical vehicles were then put into operation little by little.

The 20 vehicles move over an area of approx. 15,000 m^2 and serve over 60 stations with a total of more than 120 parking spaces. Thanks to the nickel-cadmium battery concept, a changeover to other shift models, such as three-shift operation, is possible at any time and without further changes to the system (Fig. 3.56).

The AGV master control system receives the transport orders from a specially developed PPS software, which is connected between the two WMS (warehouse management system) and the AGV master control system. Through an optimisation of the transport orders together with an intelligent priority control, the individual transports that arise can be assigned to the most suitable vehicle in each case. The vehicles plan their routes themselves, selecting the fastest and best connection between source and destination. All these measures serve to ensure that the transport orders that arise are carried out safely, reliably and, above all, at the right time. The most important point, however, is that all transports carried out are 100% traceable at all times. This is indispensable in the pharmaceutical industry.

In the solution realised here, routes, stops, transfer stations, traffic rules and much more can be created, adapted or removed by the customer himself.

3.2.9   Aviation and Supply Industry

At MTU Aero Engines in Munich, turbofan engines are produced for large Airbus and Boeing aircrafts. In keeping with the technologically demanding environment, the intralogistics solution used there is implemented with an AGV system.

Fig. 3.56 Vehicle unloaded, loaded and with load at transfer station. (Source: FROG/Oceaneering)

When you visit this special assembly line at MTU Aero Engines GmbH in Munich, you are torn between the intralogistical solution with AGVs and the assembly object, especially the end product. After all, the GP7000 and GEnx turbofan engines are used in the Airbus A380 (Megaliner) and the Boeing 787 (Dreamliner), probably the most fascinating commercial aircrafts at the moment.

Now, the engines are not completely manufactured on this assembly line, they are too large and complex for that. The finished engines have a diameter of approx. 3 m with an overall length of just under 5 m and provide a thrust of approx. 300 kN (Fig. 3.57).

The turbine centre frame, which connects the two pressure stages LP and HP, is mounted on this line. This module is called Turbine Centre Frame (TCF). At present, about 100 modules are produced annually for the GP7000 type and about 240 for the GEnx type. So seven modules are completed per week, each requiring 40 h of working time, 35 h of which are on the AGV line.

New Concept for Small Series Assembly with AGVs

At the beginning of 2010, the project team started planning the new assembly line. Due to the relatively high target quantities, a new assembly concept was decided upon. This included an automated cycle line with seven assembly stations and a pre-assembly station. To ensure a smooth assembly process, extensive prerequisites were created and implemented (Fig. 3.58).

An important goal was to minimise disruptions in assembly caused by wrong or faulty components. To achieve this goal, picking is carried out by employees who have the same level of training as the employees in assembly. This allows them to pre-check the parts directly during the picking process. The required number of components is picked with regard to the specific assembly workplace.

At the same time, the aim was to achieve a compulsory clocking of the entire line. This required an automated conveyor system. After extensive market research and comparison of the possible solutions, the decision was made against conventional conveyor technology (such as chain, roller, belt, apron conveyor, skid conveyor) and in favour of an AGV system. The main reasons that spoke in favour of the AGVS in this comparison are:

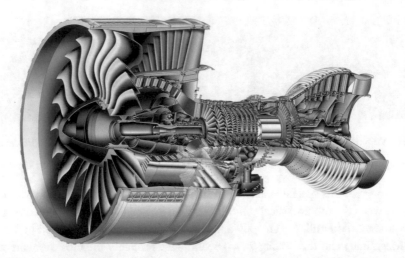

Fig. 3.57 Parts of the GP7000 turbofan engine are built on AGVs in Munich. (Source: baisi.net)

Fig. 3.58 View into the assembly line. (Source: dpm Daum+Partner Maschinenbau GmbH)

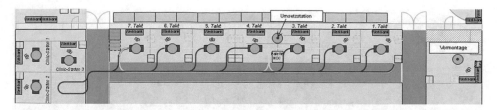

Fig. 3.59 Overview of the assembly line layout (Source: MTU Munich)

- Space-saving design,
- best accessibility to the installation object,
- high flexibility of the AGV system for easy conversion of the assembly line,
- no superstructures on the ground, thus free paths and areas,
- no cost disadvantages compared to conventional conveyor technology.

Figure 3.59 schematically shows the layout of the realised assembly line. After pre-assembly, the actual line with seven assembly stations is started. The first three and the rear four stations are each combined into individual circuits. In each of these two circuits, three or four AGVs are used, which have different receptacles depending on the assembly progress. In total, there are seven AGVs in the system, one at each assembly station.

After pre-assembly, the pre-assembled components are delivered to the respective stations of the first AGV circuit, where a defined scope of components is assembled and installed in the housing. The assembly is then transferred to the first vehicle of the second circuit (from assembly station 4) on a universal tool, and assembly continues at the remaining four stations of the second circuit. Afterwards, the TCF module is finished.

A Complex AGV Master Control System Is Not Necessary
The automatic transport system does not require an extensive control system in the classical sense. What is important is the compulsory clocking, the option to eject any single AGV at each assembly station and the system visualisation.

At 5 h, the cycle time is unusually long for a clocked assembly line. A large monitor is hung at each assembly station so that the assembly staff can see at any time how much time still remains of the cycle and how the work is progressing at each of the seven stations. If there are any faults, these are also immediately visible on all monitors. The simplest displays in the traffic light colours red, yellow and green are used, so that it is possible to see at a glance whether there are problems and, if so, where they are. Through these measures, an extremely high level of reliability and installation quality has been achieved.

If a quality issue arises at any assembly station that cannot be solved on site within 30 min, the AGV and the module are ejected in their as-built condition and moved to a separate "repair station" where a special team fixes the problem.

Fig. 3.60 The automatic transport vehicle with rotating assembly body. (Front and rear view; source: dpm Daum+Partner Maschinenbau GmbH)

Automated Guided Vehicles with Standardised Technology

The seven AGVs are characterised by uncompromisingly standardised technology and sophisticated design and thus form the substructure for the different superstructures in the two assembly circuits (Fig. 3.60).

In the first circuit, the structure is used for the gas channel of the TCF (flow path hardware). In the second circuit, the AGVs pick up a housing superstructure that allows the installation of internal parts as well as mounting from the outside. Both superstructures can be rotated for ergonomic reasons and remain on the AGV as long as there is no reason to change the assignment of the vehicles to the circuits. Incidentally, the superstructures were not included in the scope of delivery of the AGV manufacturer, but were made by MTU's jig construction department. Just the mechanical and electrical interfaces were determined together with the AGV supplier.

The vehicles have battery-powered drives. The lead-acid traction batteries are charged by means of a charger, which is installed in the vehicle and manually connected to the socket at the weekend by means of a power cable. More frequent charging is not necessary due to the long dwell times in the assembly areas and the low proportion of driving time.

The safety concept is simple but effective: All vehicles are equipped with a safety laser scanner for workers' safety & protection in the direction of travel. Apart from the obligatory emergency button as well as the flashing lights and the acoustic signals, no further safety measures are required, especially since the driving speed of 0.5 m/s is moderate.

Vehicle navigation is also solved simply and effectively. Due to the clear layout of the flow line system, the optical line guidance is completely sufficient. A black guide track is applied to the light-coloured floor, which the vehicles use for orientation. Stopping and branching points are realised with transponders embedded in the floor.

Project and Operational Experience

The system was put into operation in February 2011. For the application with the relatively simple assembly and track layout, the realised track guidance by means of a coloured tape glued to the floor is optimally suited. The tape can be easily adapted in any case and is operationally reliable in the long run. So it is quite conceivable that in near future the current seven assembly stations will become more—no big deal for the concept and the technology used.

3.2.10 Plant Engineering

On the factory premises of a plant manufacturer, an automatic heavy-duty vehicle transports machine parts weighing up to 63 tonnes between the various production halls. The 6-m-long and 2.50-m-wide platform vehicle also passes a 140-m-long section outside.

At the points on the course where man-guided vehicles cross the drive track of the AGV, a traffic light system with a half barrier for each direction of travel is installed to control the traffic. As soon as the AGV enters the crossing area, the master control system switches the traffic lights to red for the cross-traffic and the barriers are lowered. After the AGV has left the area are the traffic lights turned green again and the barriers raised.

Since no laser scanner approved for workers' safety & protection was available for outdoor use at the time the vehicle was built, the vehicle drives with radar sensors. They are mounted at the front and rear. Safety edges all around and soft bumpers at the front and rear side complete the safety devices (Fig. 3.61).

The chassis consists of a stable welded construction and four steered pendulum axles. The hydraulic steering has a wide range of adjustment so that the platform truck can be moved well even in confined spaces despite its size.

The vehicle is loaded manually with a crane at the changing stations. Here, too, there is a safety precaution: a rotating beacon indicates to the crane operator when an AGV is in the intersection area with the crane. Only after the crane has left said intersection area, i.e. is outside the vehicle contour and load, the worker can restart and release the movement of the AGV by pressing a button. The vehicle, which navigates with a magnetic grid, then heads for its destination station.

The whole transport system is controlled by am AGV master control system. On eight industrial terminals it visualizes the vehicle's current position. The workers also enter the driving orders for the AGV there.

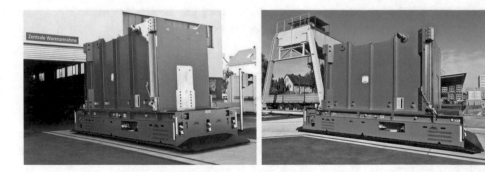

Fig. 3.61 Automatic heavy-duty transporter runs outdoors and transports loads of up to 63 tonnes. (Source: MLR)

3.2.11 Warehouse and Transport Logistics

3.2.11.1 Automatic Narrow Aisle Truck in a High Bay Warehouse

The logistics service provider DSV Solutions operates a high bay warehouse with 30,000 storage locations near Amsterdam, where an automatic transport system independently stores and retrieves the pallets loaded with cocoa.

In the high-performance warehouse, DSV uses six automated high bay stackers that move up to 160 pallets per hour. The vehicles, which are equipped with a telescopic table, can extend the forks to both sides to store and retrieve the load units at left and right side of the aisle.

Fine positioning by laser scanner ensures that the exact transfer height is maintained. During storage, the laser scanners not only measure the height above the crossbeam, but also the empty space of the storage location. If deviations have occurred due to incorrect positioning of the mast or the vehicle's base unit, the measuring equipment can be used to realise a readjustment. The laser technology works extremely fast, the measurement of the empty space takes only a few milliseconds (Fig. 3.62).

The 1.50 m wide narrow aisle trucks move loads of up to 1.3 tonnes and can extend the lift mast up to 10.5 m. When travelling in the aisles, the vehicles achieve speeds of up to 2.7 m/s. The vehicles are equipped with magnetic navigation and can independently change over to another aisle.

A particularly economical energy concept was developed for the system: One battery charge lasts for more than two shifts. Only then the vehicles have to be connected to the charger and the batteries are fully recharged for the next use.

3.2.11.2 Efficient Fulfillment with Automatic Transport Vehicles

Hermes Fulfillment GmbH operates a large shipping centre for e-commerce in Ohrdruf, Thuringia. An automatic transport system ensures fast and smooth picking and dispatch of the product ranges of the parent company OTTO and other customers.

Fig. 3.62 Automated narrow-aisle stacker for use in high bay warehouses. (Source: MLR)

The logistics location in Ohrdruf specialises in a multi-layered assortment of articles ranging from electrical appliances, smaller furniture, home accessories and carpets to DIY articles. The market goods may weigh a maximum of 31.5 kg and have a maximum edge length of 3 m.

The Order Picking Area
The total extension of the AGV drive track layout is 13 km on an area of 366 × 140 m and includes a high bay warehouse of 50,000 m². This warehouse houses the order-picking goods, which are manually removed by the order pickers on the lower three levels. The 3rd level is at a height of 4.70 m (Fig. 3.63).

Since very different, sometimes quite large packages are processed in Ohrdruf, one pallet is usually not enough for a picking order, so a 2nd and 3rd pallet are needed. If this task were to be handled with a conventional solution based on manually operated forklift trucks, the order picker would have to drive into the warehouse several times and bring out each pick-to pallet. Very long distances would have to be travelled several times, and the time required would be correspondingly high.

The new system relies on AGVs, which enables the order picker to order additional vehicles (with empty pallets) in time if needed. Thus, he picks on one vehicle and fills the pallet there, but can request another vehicle with an empty pallet in time. When the current pallet is completely filled up, he sends the AGV on an automatic drive to the goods out area and switches to the next AGV himself. In this way, he can continue to work continuously and does not lose time by driving away the full pallet and fetching an empty one (Fig. 3.64).

The AGV is monitored like the other automation components of the site in the control centre. The current positions of the total of 52 AGVs are displayed on extra-large flat

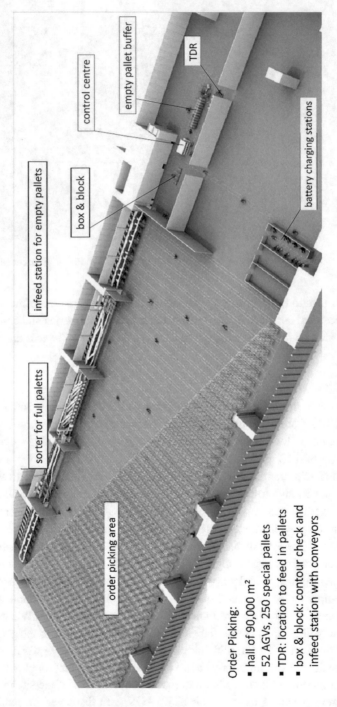

control centre

empty pallet buffer

TDR

infeed station for empty pallets

box & block

battery charging stations

sorter for full paletts

order picking area

Order Picking:
- hall of 90,000 m²
- 52 AGVs, 250 special pallets
- TDR: location to feed in pallets
- box & block: contour check and infeed station with conveyors

Fig. 3.63 Schematic overview of the system layout. (Source: DS AUTOMOTION)

Fig. 3.64 The two vehicle types in the main aisle in front of the warehouse racks. (Source: DS AUTOMOTION)

screens. Via an interface to the HERMES warehouse management system, the picking orders reach the AGV master control system and are assigned to the individual employees and vehicles there (Fig. 3.65).

The AGV master control system is connected to the AGVs via WLAN, so the AGVs receive their orders and report their status themselves.

Special Pallet and Palletising Robot

The order picking *pallet* is a special box with the dimensions of 1400×1200 mm (L × W). It is made of plastic and has foldable side walls so that it can be stacked when empty. The side walls have been reinforced with special construction elements. Since the box must be open on the side facing the storage locations during the picking process on the upper levels, there is a circumferential tensioning belt that gives the whole thing the necessary stability.

A market research did not provide a suitable product on the market. For this reason, an in-house development was started for the goods-specific requirement of the AGV process, which included the selection of the materials, the construction elements and the constructive design. Today, 250 of these boxes are in use—they remain on site and are almost indestructible.

At the empty pallet buffer, a palletising robot takes a folded empty box and places it on an AGV. Equipped in this way, the AGV moves to the aisle where the picker who has

Fig. 3.65 The control centre—not only for the AGV system: Full control in front of large flat screens. (Source: DS AUTOMOTION)

called an empty vehicle is working. The picker takes over the AGV, unfolds the box and can continue with his picking order (Fig. 3.66).

The Ride-On AGV
The most frequently used picking level is on the floor (level 1). This does not require a lift on the AGV, the pallet is held close to the ground by the pallet truck. The CX-M type AGV is available for this purpose. This is an automated version of a Still series device (horizontal order picker). There are 40 units of this variant in the plant (Fig. 3.67).

The AGV type EK-X is used for order picking in the upper levels 2 and 3. This is an automated variant of a Still series forklift (vertical order picker). There are 12 AGVs of this type of vehicle (Fig. 3.68).

In total, the AGV comprises 52 vehicles. The table shows the main technical data (Table 3.5).

There are not many AGVs that have been designed to carry people. There are a few people movers that have the purpose of transporting people—especially in the public sector. There are also automatic transport vehicles that can be operated either automatically or manually—these are mostly simple forklift trucks.

In this application, however, it is a question of vehicles on which persons travelling with the vehicle are alternately driven during automatic operation and carry out picking activities. For this purpose, two European standards had to be fulfilled at the same time, namely DIN EN ISO 3691-4 and DIN EN 1526. Usually, developers/manufacturers only have to deal with one of these two standards. The AGV manufacturer DS AUTOMOTION had to do some pioneering work and developed the solution outlined below together with external consultants and the employers' liability insurance association.

The installed safety concept for the ride-on AGV includes a footboard, safety mats and the so-called safety balls. The footboard is located in the middle of the vehicle, i.e. where the driver/worker is standing. Around the outside are the safety mats, which cause the vehicle to stop immediately if they are stepped on. This forces the employee to stay in the

Fig. 3.66 The pallet robot loads an AGV with a folded special pallet. (Source: DS AUTOMOTION)

Fig. 3.67 Riding on the AGV: Two contacts in the base plate as well as the two safety balls must be actuated during travel. (Source: DS AUTOMOTION)

Fig. 3.68 Manual order picking on the 2nd and 3rd racking level. (Source: DS AUTOMOTION)

middle of the vehicle while driving. Even if a second person were to enter the vehicle from the outside during the automatic drive, one of the safety mats on the outside would trigger and stop the drive.

During the automatic drive, the order picker must actuate a safety ball with each hand that is in front of him at gripping height. The AGV will only move when the footboard and both safety balls are actuated by the driver. When the stop position is reached, the driver releases the safety balls and can carry out his picking activity.

Table 3.5 The most important characteristics of the two vehicle types

Parameter	AGV type CX-M	AGV type EK-X
Dimensions and weights	L × W: 2942 × 940 mm Height: min. 1489—max. 2075 mm Weight: empty 1350 kg Load: 1000 mm/LSP 600 mm	L × W: 3332 × 1200 mm Height: min. 2903—max. 7000 mm Weight: empty 2900 kg Load: 800 mm/LSP 400 mm
Driving speed	Forward: 1.6 m/s Backwards: 0.3 m/s	Forward: 1.4 m/s Backwards: 0.3 m/s
Navigation	Magnetic point sequence	Magnetic point sequence
Safety	Ride-along technique, additional: Indicator, emergency stop button, load flap for pallet, bright rotating beacon for better "visibility" in the racking aisles	Ride-along technique, additional: Indicator, emergency stop button, load flap for pallet, bright rotating beacon for better "visibility" in the racking aisles
Personal safety	Safety laser scanner	Safety laser scanner
Vehicle kinematics and drive	3-wheeler with support wheel 3 kW/24 VDC	3-wheeler 3 kW/24 VDC
Lifting device	Hydraulic, from 86 to 786 mm	Hydraulic, from 65 to 5415 mm
Energy concept	Lead gel battery 24 V/450 Ah	Lead gel battery 24 V/930 Ah

Summary

The project was realised in 2011, after which the system's performance was ramped up until January 2012. Since March 2012, the system has been running without any major disruptions. The AGVs are in operation at the Ohrdruf site for 12–14 h a day. The energy concept of the vehicles is based on lead-gel traction batteries, which do not need to be charged during the day while in use. In the evening, the vehicles drive to the automatic battery charging stations, where they are charged during the night so that the entire vehicle fleet is available again with full batteries the next morning (Fig. 3.69).

3.3 Outdoor Use (Outdoor AGV)

If one wants to operate automated transport vehicles outside of factory halls, the topic becomes much more complex. Here we are still limiting ourselves to (actually) in-plant use, because we are looking at the operation of such a system on the factory premises, i.e. away from public road traffic. The term "external industrial environment" is also used for these applications.

Due to the special requirements and increased complexity compared to the previous examples, no concrete outdoor AGV projects will be presented in the following. Instead, the focus will be on explaining the special circumstances surrounding the use of outdoor

Fig. 3.69 The AGVs at the automatic battery charging stations: power is supplied via the contact plates on the floor. (Source: DS AUTOMOTION)

AGVs, which we believe will be more informative for the reader overall and ultimately more helpful in answering the question of whether such a system can also be used in their operations in the future.

The special operating conditions and the associated requirements can be roughly summarised as follows:

- Long distances require high speeds, otherwise no satisfactory transport performance can be achieved; 10 km/h is desired if possible, sometimes even more.
- Higher speeds mean longer braking distances and require correspondingly far-reaching sensor technology for the safe detection of obstacles (people!) in the driveway in front of the AGV.
- The road condition is worse than on the drive tracks in the halls; this results in different running gears and wheels.
- Different road surfaces—materials, but also wetness/snow/ice—lead to different "grip", i.e. the traction and thus the braking capacity and the braking distance change.
- Road traffic is rougher outside and the traffic situation more complex than indoors.
- All vehicle components, especially the necessary and sensitive control and sensor systems, are exposed to considerable mechanical stress as well as weather influences.

The influence of the weather on an outdoor AGV is manifold, possibly even considerable, and will therefore be considered in more detail. In detail, the following weather conditions can be found in Central Europe:

- Very high temperatures in summer (up to +45 °C),
- very low temperatures in winter (down to −30 °C),
- very large temperature changes with mixed indoor/outdoor operation,
- extremely varying light conditions (darkness, cloudy sky, extreme sunshine, high and low sun),
- fog,
- snowfall, sleet, hail,
- black ice,
- heavy rain,
- strong wind (up to gale force).

It must be clear to every user that weather conditions can become so extreme that AGV operation must be temporarily suspended. This certainly includes icing of the roads as well as extreme snowstorms: If the grip, i.e. the contact between the wheels and the ground, is no longer given, safe operation is also no longer possible. The operator is obliged to recognise such a situation and stop the operation of the AGV until regular and controllable boundary conditions exist again (Fig. 3.70).

The weather therefore has an influence on the design and construction of an outdoor AGV. Many of the relevant points are not new, however, but are commonplace, not least in commercial trucks. For two typical AGV functionalities, however, the weather poses an additional and particularly demanding challenge: Safety and navigation.

Fig. 3.70 Automatic truck in winter operation. (Source: Götting)

Both functionalities are addressed in detail here to highlight these special features, even though the basics of AGV technology have already been covered in an earlier chapter.

3.3.1   Outdoor Safety

In the system design of an outdoor AGV, safety, i.e. worker's safety & protection, is the top priority. In the following, the challenges of outdoor use will be explained, also by comparing them with the corresponding indoor conditions.

The Internal Use
The certified safety laser scanners are intended—both indoors as well as outdoors—for use in industrial environments with instructed personnel. As soon as this requirement is no longer met, they lose their approval.

If an AGV is driving on the factory premises, the traffic situation is often much more complex than inside a factory building. All kinds of road users may be on the road on the factory premises and it is extremely difficult to ensure that all road users have received instruction on how to "deal" with an encounter with the automated vehicle. If it is difficult enough to recognise, intercept and, if necessary, escort or instruct non-operating persons in the aisles, this is hardly possible outdoors.

High Speeds
In principle, it can be assumed that the speeds—of all road users—are higher outdoors than indoors. Often, 30 km/h is allowed on the factory premises, which is driven by cars, trucks with and without trailers, tractors and buses. In addition, the vehicles (manually and automatically operated) are usually heavier than indoors, so the consequences of collisions can be considerable.

It should also be noted that doubling the speed already quadruples the braking distance!

Currently, depending on the conditions, a maximum of 10 km/h is possible for AGV. But even this is already associated with increased costs, i.e. up to 6 km/h is usual. In the near future, 15 km/h will be allowed under certain conditions (e.g. level road, all wheels braked).

Note: Signs on the factory premises such as "StVO applies here" are self-imposed rules, because the StVO[18] does *not* in itself apply to internal factory traffic. If the StVO applied, automatic vehicles would also have to be registered. Only slow vehicles are exempt from the registration requirement, e.g. according to §4 FeV[19] paragraph 1.3 "tractors that are designed to be used for agricultural or forestry purposes, self-propelled working machines, forklifts and other industrial trucks, each with a maximum speed determined by their design

---

[18] German abbrevation, meaning: Road Traffic Act.
[19] German abbreviation, meaning: Driving Licence Ordinance.

of no more than 6 km/h, as well as single-axle tractors and working machines that are guided by pedestrians on bars".

A challenge for outdoor AGV is therefore the large speed difference: if the automatic vehicles are travelling at a maximum of 6–10 km/h and the other road users at up to 4 times the speed, this poses risks. It must also be considered that the automatic vehicles are easily perceived as obstacles by the drivers of the manually operated vehicles, which may lead to daring overtaking manoeuvres, stress and an increased risk of accidents. Indoors, we also know these phenomena, but mostly because of slow shunting speeds, less because of the usual indoor driving speeds of 1.0–1.7 m/s (3.6–6.1 km/h).

Reduced Friction

Due to the sometimes considerably lower friction (tyre–road), very long braking distances may result. This is caused by dirty (greasy), wet or icy roads. To keep the braking distance short even on wet or icy roads, you must brake with all wheels if possible. On dry roads, however, this can lead to excessive braking deceleration. This in turn can be problematic if the load on the AGV is not secured, which is often the case (load slips or tips over). It is therefore particularly important with outdoor vehicles to detect obstacles as far ahead as possible and to adjust the speed in time to avoid an emergency stop.

At high speed, the emergency stop area must be designed accordingly wide. When cornering, the emergency stop range must be partially reduced and, of course, the speed must also be adjusted.

Roadways

When using AGVs indoors, a lot of attention is paid to the floor (floor quality). This is because the floor is of fundamental importance for the safe and trouble-free operation of the AGV. The most influence can be exerted when the floor is newly manufactured. Then the relevant standards and guidelines can be consulted, which can be found, for example, in VDI Guideline 2510-1.

A precise description of an "AGV-compliant" floor would go too far here. Generally speaking, it is defined by compliance with certain standards of the following criteria:

- **Compressive strength** of the road surface: The high surface pressure as well as the equally high shear forces are important. It should also be noted that, in contrast to manually operated vehicles, AGVs always use the exact same tracks.
- **Friction:** The coefficient of sliding friction should be between 0.6 and 0.8. If it is lower, proper emergency braking is not guaranteed; higher values cause excessive wear on the wheels of the AGV.
- **Uphill and downhill stretches: Uphill stretches** must be controllable by the vehicle on the drive side and **downhill stretches** pose risks in the event of emergency braking—the vehicle must neither tip over nor have longer braking distances due to slipping. Sufficiently large transition radii (up to approx. 25 m) must also be ensured so that the AGV does not touch down with the frame when driving up the slope, because the ground clearance of the vehicles is only a few centimetres for safety reasons. Five to seven percent gradients are normally no problem.

- **Cleanliness:** The floors must be cleaned regularly during the operation of the AGV, making sure that the floors are completely dried after cleaning, because wet floors can lead to unsafe driving manoeuvres.

The roadways on which the AGVs run can and usually may be shared by other road users, e.g. pedestrians, cyclists, forklift trucks. They must be visually marked as such. Whether additional safety measures or facilities are required due to limited roadway widths will be clarified in the project phase with the health and safety authorities (Government Agency for Occupational Health and Safety) and the relevant Employers' Liability Insurance Association.

In most cases, indoor AGVs are unsuspended and have wheel bandages made of Vulkollan. For safety reasons, the vehicles are "lowered", i.e. the ground clearance, i.e. the distance between the frame and the floor, is only a few centimetres, so that no foot or other objects can get under the vehicle.

In outdoor areas, such strict requirements are usually impossible to meet. The unevenness of the ground is great, there are driveways, road surface changes (concrete, asphalt,...) and inclines/declines. Usually, outdoor AGVs are also larger and heavier than classic indoor vehicles and run on solid or pneumatic tyres. This results in the following difficulties for outdoor projects:

- There are surfaces that are unsuitable. The classic asphalt road surface is an example of this, because the precise tracking of the AGVs very quickly causes ruts. On summer days, the temperature of the asphalt rises up to 70 °C—the asphalt becomes soft and "flows"—a no-go for laying markers (magnets or transponders) in the ground.

In addition, the automobile club ADAC, for example, warns that there is already a danger of slipping on the asphalt from 45 °C, i.e. the coefficient of sliding friction required for short braking distances changes for the worse.

- Wet floors can be prohibited or avoided indoors, but hardly outdoors. So you have to reckon with rain-soaked roads, on which you have to drive with suitable tyres and at an appropriate speed, depending on the project.
- Snow- and ice-covered floors: Under such conditions, safe driving is impossible; therefore, the task remains to automatically or at least reliably detect such conditions, to (temporarily) stop the AGV operation and to provide and use an alternative solution for material transport.
- Large wheels, high ground clearance, steered wheels that protrude from the contours of the frame while steering and the high driving speed aimed for make safe operation difficult.

In the outdoor project, if you manage to keep the driveways and operational areas of AGVs free of people, you no longer have many problems. In such cases, no persons can come to harm, and many things are possible. However, separated tracks can rarely be realised; so workers' safety and protection is usually on the agenda.

Fig. 3.71 Safety devices on an outdoor AGV. (Source: Götting)

Fig. 3.72 Safety system for outdoor use based on laser scanner + safety controller. (Source: Götting)

Workers' Safety and Protection

Workers' safety is clearly defined for the AGVs' world: The human body and its limbs are simulated both lying and standing by means of cylindrical test specimens. These cylinders with a diameter of 70 mm and a length of 600 mm are set up vertically and thus replace the lower legs of an-adult-person. The safety device used at an AGV must detect the test bodies and bring the vehicle to a stop before the test body comes into contact with a fixed part of the vehicle (or its load).

For indoor use, non-tactile sensors based on laser scanners have been available for more than 25 years and have been approved by the Employer's Liability Insurance Association as a safety system (see also Sect. 2.1.2.4). Therefore, most indoor AGVs today have a safety laser scanner for workers' safety and protection. Tactile mechanical safety devices such as soft foam bumpers or plastic stirrups are only rarely used. In addition to the laser scanners, safety edges for the emergency stop are often used, which are mounted in the direction of travel and on the sides of the vehicle.

Until 2018, such safety laser scanners did not exist for outdoor use: On the one hand, the use of laser scanners is more complicated in all conceivable weather conditions (e.g. there is a "risk" of many false activations due to snowflakes/rain drops, whirled-up dust, birds/insects flying by, etc.), and on the other hand, the time and financial expenditure for approval as a safety system is very high—and this in a rather small market with expected low sales figures. Therefore, in the past, outdoor AGVs that do not travel in demarcated areas free of people were equipped with relatively large mechanical emergency stop bars or with voluminous soft bumpers. The laser sensors that were additionally attached merely support this mechanical bumper. On their own, they must not be responsible for safety (Fig. 3.71).

The size or length (in the direction of travel) of the mechanical bumpers is limited purely by design and in turn limits the travel speed of the AGV (bumper length = braking distance). For a max. driving speed of 6 km/h, a bumper of approx. 1.30 m length is typically required!

There are now two certified safety laser scanners from two suppliers available that may be used outdoors to safeguard an AGV: the HG G-4500 by company Götting and the outdoorScan3 by company SICK. The two systems differ in the way they are mounted on the vehicle or in the position of the scanning plane as well as in the sensor range and thus in the maximum permitted speed of the vehicle (Figs. 3.72 and 3.73).

Radar technology is much less dependent on different weather conditions. However, the angular and range resolution is poorer than with a laser sensor. Radar sensors have difficulties detecting objects that are close to the ground, such as people lying on the floor.

Ultrasound is only suitable for short ranges, otherwise there is too much interference. In particular, air movements (wind), which must always be expected in outdoor areas, falsify the measurement results. Furthermore, ultrasonic sensors, like radar sensors, have a relatively poor resolution (accuracy) in distance and direction.

It would be conceivable to achieve a satisfactory result by means of so-called sensor fusion, i.e. the combined evaluation of data from different sensor systems. The demands on a safety system designed in this way—especially on the evaluation software, which is supposed to guarantee the safe overall function under all circumstances and would, for example, also have to detect a malfunction independently—are considerable. For this reason, among others, such a safety system is currently not available on the market.

Fig. 3.73 Safety laser scanner for outdoor use. (Source: Sick)

Machine Protection

Today, non-tactile safety is provided by 2-dimensional laser scanners. However, these sensors do not prevent the AGV from hitting a solid body. Due to the low mounting height (approx. 10–15 cm above floor level) and the 2-dimensional mode of operation, these devices scan on a single plane close to the ground only.

Figure 3.74 shows examples of objects or hazardous situations that are not (or cannot) be detected by these sensors. So here it can happen that the laser scanner "looks" above or below and a collision occurs.

Outdoors, the number of obstacles that might not be detected by the 2D safety scanner increases significantly once again (Fig. 3.75).

Figure 3.76 shows sensors and measurement principles that are additionally required for 3D machine protection. None of these sensors are certified for workers' safety, so they cannot be used on the AGV alone. Therefore, to date, the sensors used for machine protection are necessary in addition to the mandatory safety sensors.

The sensors and measuring principles shown as examples in Fig. 3.76 cannot be used without problems even for indoor applications, as all sensor principles have their specific limitations and problems. The essential characteristic of all the examples shown is the need for correct parameterisation to the objects and their properties that are to be detected. There is no sensor that could detect all interfering objects under all conditions.

The typical specification of a machine protection/object protection sensor is:

- Load condition and dimension-dependent full volume monitoring of the AGV clearance profile
- often reflective materials, e.g. aluminium, chrome or VA; or transparent (e.g. glass or Plexiglas); or black
- typical test object: square 20 × 20 mm or tube 20 mm diameter

Fig. 3.74 Examples of obstacles that are difficult or impossible to detect

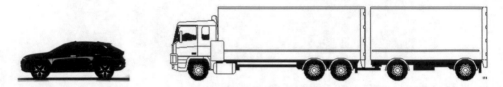

Fig. 3.75 Cars and trucks with trailers, both with high ground clearance, i.e. difficult to detect as an obstacle from the side using a 2D laser sensor close/parallel to the ground

Fig. 3.76 Alternative sensors for machine protection

If it is already problematic to find a suitable machine protection sensor for indoor use, the problem becomes even greater outdoors, if only because of the different lighting conditions. Calibrating these sensors is the big challenge: if the calibration is too coarse, the sensor may miss an approaching hazard, and if the calibration is too fine, there are always false alarms and malfunctions that cause the AGV to stop and unsettle the rest of the traffic.

Conclusion

With the sensors available on the market, it is basically possible to set up a non-tactile safety system for outdoor AGVs that complies with the applicable regulations. This means that there is no need for protruding mechanical safety guards. However, since these scanners only measure close to the ground and only capture a 2D image of the surroundings, it is urgently recommended that additional sensors for machine protection, i.e. for detecting other obstacles on the roadway, shall be mounted on the outdoor AGV. Furthermore, it is recommended to limit the driving speed to 6 km/h for the following two reasons:

- This speed limited clearly increases safety compared to today's manual practice and creates confidence.
- For speeds above 6 km/h, approval for public road traffic is required[20] which will increase the effort significantly.

3.3.2   Outdoor Navigation

In principle, there is a wealth of procedures and technical systems available on the market to navigate an AGV. However, if one is talking about outdoor use, most of the procedures are no longer an option. According to the current state of the art, the following three procedures remain for outdoor use:

- Transponder navigation (see also Sect. 2.1.1.2)
- GPS navigation (see also Sect. 2.1.1.5)
- Laser navigation (see also Sect. 2.1.1.3)

Due to its limited range and sensitivity to intense sunlight or heavy rain/snowfall, laser navigation is not suitable for purely outdoor use. However, it can be used if the vehicle is mainly driven close to hall (walls) or only short distances between halls have to be covered outdoors, i.e. if the aforementioned influencing factors do not have a negative effect.

Transponders are data carriers with unique coding; they are embedded in the ground along the drive track, usually only a few centimetres below the surface. The AGV is equipped with a transponder reading antenna on its underside. There are transponders of different sizes, which then enable different reading distances and accuracies in combination with the vehicle antenna used. In general, the smaller the distance between the antenna and the transponder, the higher the achievable accuracy. The transponder sizes vary from approx. 2–8 cm in diameter, the reading distance from approx. 10–40 cm and the achievable repeatable accuracy from approx. 2–20 mm. The installation process is very simple if the road surface is free of metallic components. If this is not the case, the transponders must be inserted into significantly larger holes and the space in between must be filled with concrete or synthetic resin, which increases the installation effort accordingly.

This type of navigation is very robust and reliable, but it requires a solid ground/road surface that is just as unimpressed by the midday heat in midsummer as it is by heavy vehicles driving over it. Also to be noted are the limitations due to rather low ground clearance of the vehicles and certain restrictions of flexibility in case of necessary changes of the drive track layout. The automated trucks shown in Figs. 3.70 and 3.71 use this type of navigation.

---

[20]The Vehicle Registration Ordinance 5 (German: FZV), which came into force on 01.03.2007, regulates the registration of motor vehicles over 6 km/h maximum design speed (German: bbH) and their trailers.

Another method or technology that is suitable to navigate AGVs outdoors is the so-called Real Time Kinematic Differential GPS (RTK dGPS).[21]

Unlike transponder navigation, this navigation technique works on any ground, but needs a clear view of the sky (to the satellites). High walls close to the AGV travel path are a disadvantage, as are bridges or other travel path crossings, e.g. pipelines. Typically, the system needs a clear cone of sight of about 15° upwards from horizon to work reliably.

The steps to achieve the required driving and positioning accuracy are:

1. Checking the local conditions, especially the reception strengths of the satellites
2. Use of the differential GPS
3. Real Time Kinematic Differential GPS

The achievable positioning accuracy of the "normal" GPS is about ±12 m. By setting up a reference station whose position is precisely measured, a correction value can be transmitted to the AGV via a short-wave transmitter. This can improve the accuracy to about ±1 m (differential GPS). If, in addition, a real-time evaluation of the carrier phase of the received satellite signals is carried out in the mobile receiver, an accuracy in the small centimetre range can be achieved. The outdoor AGV shown in Fig. 3.77—in conjunction with high-quality drive technology and algorithms for path control—achieves a repeatable accuracy in vehicle positioning of ±15 mm.

Since there are now several satellite positioning systems available (apart from the American GPS also Glonass, Beidou and Galileo), the general availability of satellite positioning has increased significantly. Where the view of the sky is too limited, transponders or even magnets can be used as alternative or supplementary waypoints ("hybrid navigation"). This technology is suitable for all weather conditions, both indoors and outdoors.

In the meantime, optical systems have also been further developed and have become interesting for some outdoor applications. It is important that the line of sight to fixed landmarks is given—even in rain, snow and fog. "Good landmarks" must be sufficiently available, they must offer clear contrasts and must not be obscured by vehicles, storage facilities or other objects.

Conclusion

With the navigation systems available on the market today, a way can always be found to reliably guide outdoor AGVs. It should be noted in this context that due to the high achievable accuracy of the navigation procedures, the AGVs—in contrast to manually operated vehicles—always follow the exact same lanes. However, if the pressure resistance

---

[21] GPS = Global Positioning System, officially NAVSTAR GPS, is a globally available satellite system for position determination and time measurement operated by the US Department of Defence; use is possible free of licence/fees.

Fig. 3.77 A heavy-load outdoor AGV (one-off design) moves packages of concrete stones from conveyor belts to a staging area; navigation is done based on RTK dGPS. (Source: Fraunhofer IML and Götting)

of the road surface is not sufficiently high, ruts can form very quickly. This is especially true when outside temperatures are high and asphalt is the road surface, which can reach temperatures of up to 70 °C in summer and then no longer has sufficient strength.

3.3.3 Summary

The current state of technology does not (yet) allow AGVs to move freely in (factory) road traffic in a similar way to driver-operated vehicles. Even AGVs equipped with safety laser scanners tend to move "blindly" and "groping", so that organisational aids are unavoidable. This "organised order" includes the following points:

- Wherever possible, driveways should be defined that are only used by the AGV. If these routes can even be mechanically separated from other traffic, e.g. by fences or other barriers, it is optimal.

- In any case, traffic areas must be clearly defined in terms of their intended use and visually marked. In this way, lanes, parking spaces, parking areas, and closed areas are clearly demarcated from each other.
- If possible, the traffic volume in the area of the AGV routes should be limited, e.g. by forming one-way streets or by restricting transit traffic.
- If the AGV crosses lanes of other traffic, a traffic light system, if possible with a barrier, is to be provided.
- Large pedestrian flows should not directly cross the drive track of the AGVs, but should be guided across the AGV's route, e.g. with a pedestrian bridge.

The main reason for these measures is that the AGV will stop too often or at least will drive slowly if it encounters a lot of cross and oncoming traffic on its routes due to the technically induced excessive consideration. This reduces the average driving speed and the transport performance.

Now, we do not want to give the impression that reliable and economical AGV use is not possible outdoors. But outdoor projects always require a serious approach with competent partners. The planning phase plays a particularly important role because it is here that the course is set for a successful project (see also Chap. 5).

In order to be able to assess the importance and feasibility of the AGV in the outdoor area, the following facts may help:

- There are relatively few outdoor projects.
- Unfortunately, there have been some AGV projects in the past that were offered by inexperienced providers and caused problems during realisation.
- There are no standards for safety and navigation as there are for indoor areas, see also the information in Sect. 2.1.

Ultimately, every new outdoor assignment must be taken seriously as a project. And every potential provider must be asked whether they have the competence and experience for such projects.

The Future of the AGV

4

Summary

We have talked about the technology and applications of the previous AGVs in the last two chapters. However, the world of AGVs is in motion more than ever, which will have a massive impact on the development of technology and markets. It would be unreliable or go beyond the scope of this publication to give a quantitative outlook on the development of the markets, which is why we refrain from doing so.

The basis for the technological changes that are now emerging lies in new, inexpensive and intelligent sensor systems as well as in software developments driven by the internet that evaluate the data from these sensor systems.

The Internet has something to do with these developments in that a lot of the groundwork in image recognition comes from Internet applications. Google, for example, has hundreds of patents on these topics. There, of course, one does not necessarily want to recognise a palette—which is a basic task for AGVs—but objects, persons (man/woman), faces up to complex situations (city, beach, forest, etc.). The first examples in mobile systems can also be found in cars. In the BMW 7 series, for example, the functions "Speed Limit Info" and "Night Vision" were available for the first time, with which speed limits (on signs at the roadside and above the road) and people (at night with thermal imaging cameras) could be detected.[1]

Today's cars do things that seemed unthinkable 10 years ago: We are all now familiar with VW's parking assistant ("Park Assist"),[2] which autonomously parks in longitudinal and transverse parking spaces. At Audi, we know the "Lane Assist" lane departure warning

---

[1] Source: BMW 2010.

[2] Source: Volkswagen 2010.

© Springer Fachmedien Wiesbaden GmbH, part of Springer Nature 2023
G. Ullrich, T. Albrecht, *Automated Guided Vehicle Systems*,
https://doi.org/10.1007/978-3-658-35387-2_4

system, which detects whether the vehicle is unintentionally leaving its lane and countersteers continuously and smoothly. The automatic distance control (Adaptive Cruise Control ACC) helps the driver in stop-and-go traffic. A sensor monitors a wide area in front of the vehicle, maintains a pre-set speed and takes into account the traffic ahead, in the range from 0 to 250 km/h.

We deliberately avoid naming and describing the sensor and control systems used here. The car manufacturers use different combinations of laser, radar,[3] lidar,[4] infrared, ultrasound and video systems. Which technologies will prevail or whether there will be combinations remains to be seen (Fig. 4.1).

In any case, modern passenger cars are now able to move partially automatically, even at high speeds. Of course, car developers undoubtedly have the advantage of being able to design functions for safety without having to take responsibility for safety. After all, the driver is still ultimately responsible in the car—and that will probably remain the case for quite some time.[5]

The different areas of automated or autonomous mobility will mix, or at least overlap, with industrial automation. Thus, requirements that are mandatory for the autonomous car or for the use of STS in public areas will probably find their way into the AGV world sooner or later. Some functions will prove useful for the AGV, others will be expected by users at some point because they are known from other application areas and have become established there.

One of the main reasons for the boom in the AGV world is the excellent suitability of the AGV in the endeavour of a comprehensive digitalisation of industrial production, which, under the keyword Industry 4.0, has also encompassed intralogistics since 2012. The AGV is an organisational tool that moves both material and data. It networks production and logistics areas, both physically and in terms of data technology. Thus, hardly any manufacturing company can avoid automated material flow with AGVs as soon as it deals with Industry 4.0.

We will now look at the functional challenges for the AGV of the future. They arise from the experiences with today's systems. More is expected in the future, forcing us to break new ground. This are not all new—we had already addressed many of them in the previous edition of the guide to AGV systems, e.g.:

- Acting quickly: The AGV still seems too slow in many applications. This applies to driving in intersection areas, communication with peripheral equipment and load

---

[3] RADAR = radiowave detection and ranging, uses radio waves.

[4] LIDAR = light detection and ranging, uses laser beams.

[5] Ullrich, G.: *Technology comparison between AGVs and autonomous cars. The importance of automation.* Technical lecture at the 26th German Material Flow Congress 2017, on 6 April 2017 in Garching/Munich.

Fig. 4.1 Driver assistance systems at Audi, Toyota and Volkswagen; top left (**a** and **b**): Audi active lane assist; bottom left: Person and ball detection in the Toyota; top right: speed limit warning system from Bosch; bottom right: parking assistant in the VW Touran. The boom began around 2010, i.e. at the same time as the 4th epoch of AGV systems

handling. In mixed operation with manual forklifts, AGVs are often perceived as obstacles.

- Truck loading: Independent measurement of the loading space and optimised loading. The first solutions already exist, but they are still too inflexible and   compared to truck loading by a manually operated forklift—too slow.
- Pallet finder: General function to be able to pick up a loading aid although it is not optimally positioned.

- Innovative operation (voice and gesture control): The operator must be recognised and his commands understood: "Wait!", "Where are you going?", "Take this pallet to the warehouse!".
- Predictive maintenance: Components monitor themselves and report maintenance/inspection needs.

One or the other AGV manufacturer will object that this or that has already been realised. That is certainly correct and commendable—nevertheless, there is still enough room to make these functions the standard for the future, safely, quickly and reliably. In the following, we will deal with the four most important functional challenges under the headings:

1. Standardisation of AGV master control systems
2. The world is not a disc
3. Drive Safe—The integration of navigation and safety
4. The autonomous AGV—How much autonomy does the application need?

4.1   Standardisation of the AGV Master Control System

The first call for more standardisation came from the car manufacturers—in the mid-1980s! The intensive use of AGVs in car factories in the 1970s and 1980s also brought problems with it: the AGV solutions all resulted from an AGV manufacturer-related project business. The AGV manufacturer brought proprietary solutions to the user with its products and know-how, which led to dependence on the supplier and high follow-up costs.

The car manufacturers Mercedes Benz, Volkswagen and BMW were part of the group that ensured that the AGV working group was founded in the Association of German Engineers (VDI) in 1986/1987. During the first 30 years, the VDI expert committee on AGV systems grew steadily and achieved a great deal: Numerous VDI Guidelines were created, and since 1991 the AGV Systems Symposium has been held regularly. For the members from the ranks of manufacturers and users, a major reason for participation was always networking. Where else could one find so many interesting facts from the world of AGV systems and authoritative people as at the meetings of the VDI expert committee?

During this long period, the goal of standardisation has hardly been realised. The guidelines have strengthened the common understanding of AGV systems' technology, but standardisation in the field of AGV master control system or of vehicles has not occurred. The project character of the AGV systems work has remained and the solutions are still proprietary. During this time, detailed knowledge about AGVS has been focussed on the suppliers of AGV systems; only very few users, planners or research institutes/teaching institutes were and are in a position to understand AGV system solutions as a whole in the same way as the manufacturers do.

Thirty years later, the call for standardisation is being heard again, and again from the major car manufacturers. At the panel discussion of the 2018 AGV Systems Symposium,

Fig. 4.2 The panel discussion at the AGV Systems symposium on 26th September 2018. Participants: Left: Andreas Forster, MLR System GmbH (Ludwigsburg)—Head of Project Planning and Project Handling and Dr Hubertus Wabnitz, E&K Automation GmbH (Rosengarten)—Head of Project Planning and Project Handling. Middle: Dr.-Ing. Günter Ullrich, Forum-FTS GmbH (Voerde)—moderator. Right: Eugen Vogt, Daimler AG (Böblingen)—Head of AGV Strategy Mercedes-Benz-Cars and Stefan David, Siemens AG (Nuremberg)—Global Account Manager Intralogistics

the background was made publicly clear: Huge new challenges are coming for the automotive industry. Eugen Vogt from Daimler AG stated that future AGV projects will be characterised by incomplete specifications and the largest AGV fleets (Fig. 4.2).

The structural change from conventional cars with combustion engines to the new e-cars means deep cuts in production technology and production logistics. Electric cars have a lower vertical range of manufacture (Fig. 4.3) and require new structures in production logistics. The uncertain future also demands maximum flexibility in terms of products and production numbers. The use of AGVs is therefore a must, because only with them can intralogistics planning be adapted to constantly changing target figures.

The correspondingly large AGV systems can no longer be supplied by a single manufacturer of AGV systems alone because it lacks the capacities, but also the diversity of supply. In a production hall, therefore, the AGV products of different manufacturers will be operated—which means that not every AGV manufacturer can use its own proprietary AGV master control system. Because that would mean that several AGV master control systems would have to be docked to a superimposed ERP system, which would mean a considerable effort. In addition, the AGVs and the AGV master control systems would have to share one common layout, which would lead to complicated and performance-consuming system interfaces.

Fig. 4.3 Less vertical integration in the electric cars of the future; here the modular electric kit (MEB) from Volkswagen. (Source: Volkswagen)

The Smart Factory, as it has been propagated as a vision of the future by Audi since around 2016, is characterised by extreme flexibility and, without exception, mobile systems (Fig. 4.4). The question will then be whether an AGV master control system, as it has been known up to now, is still sufficient for such a concept.

The vision of the automotive industry is thus a standardized AGV master control system that will free from the dependence on a single manufacturer of AGV systems and mean a huge choice of vehicles. The vehicles could be purchased from anywhere in the world, independent of the existing AGV master control system, and put into operation in a straightforward manner, thus achieving a previously unknown flexibility.[6] Figure 4.5 shows a structure of how something like this could look.

The basis of the AGV master control system are the digital maps of the Smart Factory as shown in Fig. 4.6. They contain all the information that is shared by the various mobile units (mobile platforms, mobile robots, STS, AGV).

One of the most important system interfaces is the one between the AGV master control system and the AGVs. Standardisation is not easy because there are the most diverse forms of intelligence distribution, due to the differently intelligent vehicles as well as the philosophy of the manufacturers. In 2018, a new specialist department was set up at the

---

[6] Ullrich, G., Osterhoff, W.: *AGV Systems with compatible interfaces for tomorrow*. Technical lecture on the occasion of the 7th Technology Forum "Automated Guided Vehicles (AGVs) and Mobile Robots—Opportunity, Technology, Economy" at the Fraunhofer Institute for Manufacturing Engineering and Automation on 20th September 2017 in Stuttgart. Proceedings Automated Guided Vehicles (AGVs) and Mobile Robots, Fraunhofer IPA F335, Technology Forum 20th September 2017.

Fig. 4.4 Smart factory according to Audi: Mobile, driverless production islands and manufacturing machines as well as mobile robots and drones for material delivery. (Source: Audi)

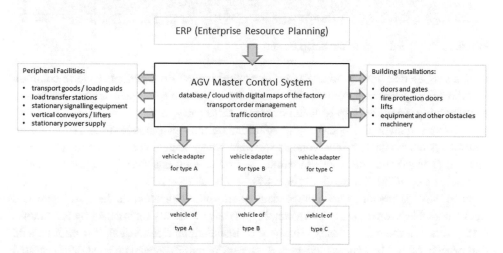

Fig. 4.5 System concept of a standard AGV master control system with three vehicle adapters for different complex/intelligent AGV types

VDMA to define a standard interface, among other things; the first version of an interface description was published in August 2019.

There are also reservations about the idea of a standard interface; these are summarised in Table 4.1. For the sake of the matter, a holistic solution to the problem must be found. The goal must be to preserve the knowledge that exists today exclusively among suppliers of AGV systems with regard to a functioning AGV system realisation (safety and performance of the system), so that the transfer of the industry into a future of whatever kind works.

Of central importance is where the sought-after standardized AGV master control system will come from. Will the big car manufacturers build their own solutions that they want to use and specify throughout the group? Will there be companies that rely

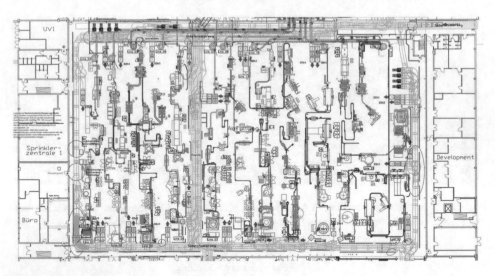

Fig. 4.6 The digital maps of the Smart Factory

exclusively on their own product[7] and offer it worldwide? Both variants have in common that the AGV system suppliers would be degraded to vehicle manufacturers and the overall responsibility would have to be taken over by a third party.

A third variant is the open source solution, i.e. an open standardized AGV master control system available to everyone as software whose source code is public and can be viewed, changed and used by third parties. A first attempt has already been made and is available as openTCS® from 2005 (Fig. 4.7).[8]

openTCS® is manufacturer-independent, platform-independent as Java software and freely available as open source software; the system is currently maintained by Fraunhofer IML. Although there are already a number of installations (references), it cannot be said that openTCS® has become established. A successful public appearance that demonstrated the functions and performance of the software was the "AGV ballet" at a joint stand at the Hannover Messe Industrie in 2009 (Fig. 4.8).

Open source software generally has decisive advantages over all other variants:

AGV system suppliers that have so far used their own solution could use this as a second pillar or use it as the main master controller, so that they could forego the expenses for further or new developments of their solution. AGV manufacturers that have so far only offered vehicles could act as a "complete" AGV system supplier.

---

[7] The first examples are: Synaos GmbH, Hanover (www.synaos.com) and GS Fleetcontrol GmbH, Lehrte (www.gsfleetcontrol.com).

[8] OpenTCS® is the result of the FAHRLOS research project funded by the BMWA—Federal Ministry of Economics and Labour, which ran from 2003 to 2005; openTCS® stands for Open Transportation Control System (www.openTCS.org).

Where industry leaders go to
synchronize their chaos.

SYNAOS

Table 4.1 Significance of a standardized AGV master control system for suppliers of AGV systems

Challenges from the perspective of the providers
– The typical AGVS project used to include the technical design, delivery and assembly of the overall AGV system. This included responsibility for the safety and performance of the intralogistics solution. The supplier was a competent system partner—this competence would be dispensed with in the future
– If in future the AGVS components are purchased individually, SOMEONE will have to take on the role of an integrator and "take the rap". The AGV system supplier will not do this any more than the programmer of the standardized master control system. This role has been unfilled until now
– AGV System suppliers need to redefine their role
– Horror vision: Decline of the AGV System culture and use of cheap AGVs from China

Fig. 4.7 The first and so far only open source variant of an AGV master control system: openTCS®

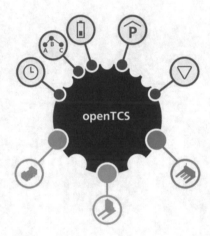

All users can use the open source software and adapt, change and develop it themselves, i.e. tailor it to their needs. They can use it as if it were their own: they can use it to plan, offer and commit to system services. In view of these serious advantages, the question arises as to why openTCS® or another open source master control system has not yet become established? Probably because the pressure from the large AGV system users was not yet great enough, but has only been strongly felt since 2017.

4.2   The World Is Not a Disc

The next sections will deal with the central question of intelligent behaviour of AGVs[9]:

---

[9]Ullrich, G.: *Der steinige Weg zur autonomen Mobilität in und außerhalb der Industrie.* Technical lecture on the occasion of the 4th Technology Forum "Automated Guided Vehicles (AGV) and Mobile Robots—Opportunity, Technology, Economy" at the Fraunhofer Institute for Manufacturing Engineering and Automation IPA on 17th May 2011 in Stuttgart.

Fig. 4.8 AGVs from seven different manufacturers under the AGV master control system openTCS® in the "AGV ballet" at the Hanover Fair 2009. (Source: Forum-FTS)

- What are the technical requirements?
- What functionalities does one expect from this?
- What can we expect from autonomous AGVs?

The technical basis for any intelligent behaviour of AGVs is sensor technology. Sensors provide the data from the environment that is necessary for the AGV to act appropriately.

Laser scanners from company Sick have been in use for the protection of persons for about 20 years, initially the PLS product line, then the S300/S3000 types and today the microScan3 product line.[10] The senor can detect persons (and any other obstacle) on the roadway ahead of the AGV at an early stage, and by software parameterization of the so called warning field it can trigger a speed reduction of the vehicle. If persons approach the vehicle up to the safety field, the sensor forces the vehicle to stop.

These safety laser scanners have become established—almost all AGVs now use them for personal safety. With this, the AGV manufacturers have fulfilled the essential legal requirements for the safe operation of AGVs within the scope of the Machinery Directive, as far as in-plant use is concerned. As important as they have been for the development of the AGV, they were and still are 2D sensors that only scan in one plane, i.e. provide the distance information of the scanning plane at installation height.[11]

To understand the problematic nature of the issue, it helps to imagine that when using this sensor technology, the environment into which the AGV drives is represented by the depth information of exactly one layer. This does ensure that the prescribed test bodies of 70 mm diameter (representing the lower legs of a worker) are reliably detected so that a

---

[10]There are other suppliers for these sensors, such as Leuze and Schmersal, but Sick was and is the market leader.

[11]The installation height usually is between 10 and 15 cm above the floor.

Fig. 4.9 2-dimensional safety sensors are not sufficient for machine or object protection

collision with the AGV is avoided. However, this information is not sufficient for intelligent behaviour of the AGV.

Figure 4.9 shows examples of obstacles in front of an AGV that are not detected if only the aforementioned safety sensors are used. These scan in exactly one plane and only check there whether the view is clear. For example, it can happen that a supposedly safe AGV drives into the side of an awkwardly parked car or van because the laser scanner "looks" between the wheels below the underfloor of the car and reports "The roadway is clear!" to the vehicle control.

In the absence of approved 3D sensors for personal safety, additional non-safe sensors are increasingly being installed for machine/object protection. Initially, they only have the task of avoiding collisions. Such sensors can also work two-dimensionally if they do not scan parallel to the ground but, for example, "look down" at an angle from above in the direction of travel and can thus detect objects above the scanning plane of the personal protection scanner. Or they are 3D sensors (e.g. ultrasound, radar, TOF cameras or multi-layer laser scanners) that scan the entire volume in front of the AGV (including the load) for obstacles.

These additional machine/object protection sensors usually do no more than warn of the existence of obstacles and cause the AGV to stop in time. The machine/object protection should—if possible—be required in every specification; however, a truly intelligent behaviour of the AGV is not yet achieved with this.

In modern AGV system applications, one can find the first steps towards multiple use of the sensors (personal safety sensor and machine/object protection sensor): For example, the measurement data of the safety laser scanner is also used for contour navigation for localisation and determining the vehicle's position. And the machine/object protection sensor can be programmed to measure the pallet to be picked up, i.e. to determine the x/y coordinates and the orientation (heading) of the pallet relative to the AGV. Hereby, a pallet can be picked up successfully and error-free even if it has been improperly put down.

It must be clear that these examples are important functions for AGV operation, but truly intelligent action is a long way off. In order to illustrate what humans can do as a matter of course with their eyes (optimal 3D sensors) and the downstream sensor data evaluation

Fig. 4.10 Recognising and verifying objects using the example of a "seating group"

within a few milliseconds, imagine a room that the human enters on one side and is supposed to leave on the other side. In the room there is a seating group consisting of a table with four chairs (Fig. 4.10).

The human recognises the situation immediately and knows how to get past the table and chairs without colliding. He can even clear his way very easily by moving one or more chairs. The vintage AGV with the common 2D sensor technology with a measuring plane 15 cm above and parallel to the floor cannot do that by far. After all, the world is not a disc![12]

If we demand intelligent behaviour from the AGV, then we need 3D sensor systems, preferably certified as safety sensor systems and equipped with complex software. There are many approaches to solving this problem, but none that have become accepted as a standard. Primarily, it is about the ability of AGVs to recognise not only people (based on the lower legs), but also objects such as crane hooks or lifted forks of a forklift truck. Automated guided vehicles equipped with 3D sensors then offer a significant increase in safety.

In future, vehicles will not be equipped with just one sensor. Several sensors of different technologies will be installed, which individually would not be sufficient, but together as fused 3D sensor systems offer undreamt-of possibilities. Why is the AGVS industry having such a hard time with this? Well, even the automotive industry uses a wide variety of sensors for the assistance functions of its cars, but the whole thing doesn't look really fused yet, although incomparably huge development budgets are available.

---

[12] Ullrich, G.: *The world is not a disc*. Technical lecture at the logistics seminar "Production Supply of the Future" on 12th October 2017 at the TU Munich in Garching.

Fig. 4.11   A Drive Safe scene with STS-to-I communication

The technical consequences of the fusion of 3D sensor data are enormous: The generated data volumes explode (compared to the data of a single 2D scanner), the data processing onboard the AGV requires much more powerful vehicle controllers and the data transmission with WiFi is no longer sufficient. The new data transmission standard 5G would certainly be a key here. Once these challenges have been mastered, however, the main task still remains, namely to develop a completely new level of software that not only evaluates this wealth of data, but can also recognise objects and patterns from it and react to them adequately. Only then one could speak of intelligent AGVs.

Safety sensors do not necessarily have to be placed in/on the AGV. There will be cases in which active support by the infrastructure makes sense (Fig. 4.11). Sensor systems of whatever kind on the ceilings of the aisles to be traversed collect information about people and other obstacles in the aisle and transmit it wirelessly to the STS/AGV. The vehicle does not need to detect all movements of people itself, but is supported by the infrastructure (STS-to-I). Such a system is particularly advantageous if it is already known what the traffic situation will be like on the new aisle before turning at an intersection.[13]

---

[13] STS-to-I : Communication between STS/AGV and infrastructure.

Fig. 4.12 Current 3D sensors for AGV use; left MRS 1000 (indoor/outdoor), right Visionary-T (indoor). (Source: Sick)

Both future 3D scanners and the novel STS-to-I communication will give decisive impetus not only to safety technology but also to navigation and form the basis for intelligent behaviour.

The sensor manufacturer Sick has two 3D scanners in its product range, neither of which is approved for personal safety, but which are certainly suitable for machine protection and further environmental detection (Fig. 4.12). On the left is the MRS 1000, a 3D lidar sensor, i.e. a multi-layer laser scanner. It has four spread scanning systems and can thus scan the volume in the direction of travel of an AGV with sufficient accuracy for many tasks. The device on the right is the Visionary-T, a time-of-flight camera, also called a snapshot camera, which takes 50 images per second at 144 × 176 pixels.

4.3 Drive Safe: The Integration of Navigation and Safety

In Fig. 4.13 of an AGV from the early days (first epoch) you can clearly see the components for navigation (finger in a groove in the floor) and safety (wire frame as safety bar). In the next epochs, there were probably further developments in this respect, but it still largely remains the classic parallel/separate construction of the two functions. In this context, classic means, for example, a magnetic sensor bar for navigation and a laser scanner for personal protection. Many years ago, we already criticised this separate processing of functions as an obstacle on the way to an intelligent AGV and called for the integration of the navigation and safety functions under the keyword *Drive Safe* as a central intelligence service.[14]

In conventional AGVs, navigation ultimately continues until a safety sensor commands a stop. The AGV operates in the world of blind robotics, just like the blind person who explores the immediate surroundings with his white cane. So, to date, we have hardly any

---

[14]Ullrich, G.: *Drive Safe—Sicher navigieren mit automatischen Fahrzeugen.* 17th German Material Flow Congress 2008, VDI Society for Materials Handling Material Flow Logistics, VDI Reports 2008, pp. 197–205. ISBN 978-3-18-092008-5. Also presentation during the conference on 4 April 2008.

Fig. 4.13 Tractor from the 1960s. (Source: E&K); analogy with the blind man's white cane

perception of the outside world. The AGV merely works through programmed sequences based on largely internal sensors. Hopefully, the reader will not perceive these lines as defamation of the AGV; in the previous chapters it could be clearly shown that today's automated guided vehicles are reliable, safe and efficient—but this chapter will deal with future requirements that are demanded especially outside of intralogistics applications.

The 2D laser scanner has become very important in the world of AGVs. Almost all new AGVs have at least one such personal safety scanner. This technology is proven and approved by the German Social Accident (German acronym: DGUV). Nevertheless: scanning just one single cutting plane is and remains unsatisfactory. Loads that protrude above this measurement plane into the roadway of the vehicle will not (cannot) be detected. An underrun AGV in the hospital manages "without any problems" to drive under the transverse hospital bed without the scanner reporting an obstacle. However, this is only problem-free as long as the AGV is not carrying a load!

Let us understand *Drive Safe* as a central intelligence achievement, namely to master situations in which the AGV master control system cannot help:

- Recognise obstacles and react accordingly—Object recognition and reaction: drive around, clear away, report.
- Recognise dangerous situations and react correctly—stop, move to the side or back, calculate alternative route.
- Detect, eliminate or prevent malfunctions—correct docking position, approach alternative docking station, report, "Pallet Finder".
- Learn new tasks quickly—commissioning by reconnaissance trip, layout changes (scheduled or unscheduled), add new AGV(s) to the system.
- Learning from repetition—adapting behaviour and movements, truck loading, space requirements in the block store.

- Participate responsibly in flowing traffic (indoor and outdoor)—adjust speed, observe right-of-way rules, "standing bearing".
- Recognise the operator and accept his voice commands—"Wait!", "Where are you going?", "Take me to …!".

Drive Safe ultimately means nothing else but the opening of the eyes. Let us imagine what it would mean for the blind person if he were suddenly able to see. He could do without the white cane and react to his environment in a completely different way. Let's use another image for the same situation:

"Constant Bearing" for Automated Transport Vehicles

Some of us are familiar with constant bearings from aviation and seafaring. When hobby sailors obtain their recreational boat licence, for example, they learn to judge the danger of a collision by looking at the course of other ships: If the approaching ship

- "eats the horizon", it will pass in front of your own ship,
- "spits out the horizon", it will pass behind your own ship.

However, if the bearing between the ship and the horizon "stands still", i.e. it does not change, then both ships are on a collision course! What does this have to do with AGV (vehicles)? An AGV is driving and navigating. To navigate means the combination of dead reckoning and bearing.

How does an AGV react in case if constant bearing, i.e. on a collision course with another moving object? This other moving object can be a forklift truck, a worker on foot or on a bicycle, or anything completely different—it doesn't matter, because today's AGVs don't even notice the impending danger of collision. Today's AGVs continue to navigate stubbornly until the safety sensor responds. Then it says "stop" and the danger is averted.

This procedure works safely to a large extent and was/is sufficient for "orderly" conditions in production areas where only adult and instructed employees move around. However, true to the motto "You are human, you think ahead!", this behaviour is not sufficient in daily life. If we are walking through the pedestrian zone and notice that another pedestrian with "constant bearing" is moving across from us, we will quickly seek an evasive strategy to escape the impending collision. We either walk a little faster or slower or change our own course to the left or right.

This is easy for us humans, but impossible for today's AGVs. But why should an AGV be able to do this? Well, because the areas in which AGVs are used are changing. The use of AGVs in hospitals has been common for 20 years, but now the vehicles no longer only drive in the secluded basements, but across the entrance hall of the clinic or along the corridors of the hospital wards. Here, the AGVs no longer encounter only instructed staff, but completely uninitiated people, and especially sick and disabled people, playing toddlers or rushing nursing staff.

Or let's think of the increasing number of outdoor operations. Automated vehicles suddenly move around on the factory premises, sometimes in the midst of people and truck drivers who are not part of the company. The preventive intelligent handling of the constant bearing method is representative of a new generation of intelligence that we want to call *Drive Safe* and ultimately means an integration of the AGV functionalities "navigation" and "safety" through novel, fused sensor systems.

Figuratively speaking, however, "constant bearing" also has a meaning for the entire AGV sector. If we imagine this as a medium-sized ship on its course across the seas, then there is an ocean liner that is also moving swiftly—and with a constant bearing! This refers to the automotive industry, which is doing a lot of development work with its driver assistance systems. Automatic functions that would look good on an AGV are available for every mid-size sedan: Automatic parking, platooning, speed & distance keeping, lane keeping, intelligent night vision and traffic sign recognition; this includes the modern sensor systems and evaluation software that we also need in the AGV of the fourth epoch.

The AGV sector—to stay with the metaphor—urgently needs to correct its course or pick up speed in order not to be overrun by the ocean liner (car manufacturers) or the giant wave (OEMs) that runs with it! This means that the AGV sector must actively face the demands of the future. It must once again question itself; it must analyse the influences from outside as well as the requirements and needs of future customers. Ultimately, it is a matter of ensuring independence with clearly defined core competencies of its own.

4.4   The Autonomous AGV: How Much Autonomy Does the Application Need?

The markets are in extreme motion: New suppliers are entering the market, bringing innovative ideas and techniques with them. This is not always a good thing. More and more often, the lack of experience and expertise is concealed by typical American marketing. Services and products are marketed highly professionally and aggressively. Beautiful pictures and "sexy" terms and phrases are used in clever advertising. Things are quickly exaggerated, claimed and promises made; the internet ist full of them.

This includes the hyperinflation of terms, especially regarding the terms "autonomous" and "autonomy". Many potential customers are asking for autonomous AGVs without really knowing exactly what that is or whether they need such a thing at all. Far too many publications in the trade press talk about autonomous vehicles. Some AGV manufacturers feel compelled to use such phrases as well in order to become or remain interesting in the market.

So the questions are:

- What does the term *autonomy* mean?
- What are autonomous functions?
- Who needs the autonomous AGV?

4.4.1   Autonomous AGV in Intralogistics

Autonomy means freedom of decision and action, i.e. a kind of self-determination or independency. Independency, however is different from autonomy. For us, autonomy is the automatic, driverless thing about the AGV. That has always been the case. Automated guided vehicles are designed to drive **automatically**, where safety does not depend on a driver. They move along predefined drive tracks with predefined tasks and are controlled and organized by a master control system in their interaction with other AGVs, other road users and the adjacent machines. AGVs have been around since the 1960s, and have always been automatic, driverless floor-based transport systems, predominantly in in-plant environments.

This means that we were and still are in the intralogistics sector with the main tasks:

- AGVs for transporting goods between warehouses and production sites
- AGVs in the warehouse either for order picking or for operating a block warehouse
- AGVs as assembly platforms in assembly lines

These tasks are carried out within the plant, usually within the drive tracks' network of a factory hall and within the scope of the Machinery Directive. Here, there are clear specifications and boundary conditions for the use of AGVs, especially with regard to the reliability of the transports, predictability, performance, availability, order, cleanliness, clear interfaces, and instructed personnel (as a pre-condition for in-house use). As long as a production or production logistics runs under these conditions, the existing AGV system is just right: a flexible, automatic, independent, reliable and economical transport system.

In intralogistics, autonomous functions are certainly being considered. And there are very real cases of exaggerated demands from technology-enthusiastic customers. In today's structures of intralogistics, however, such demands regarding autonomy functions are mostly unnecessary, counterproductive and lead to chaos. For example:

- The autonomous AGV is able to detect an obstacle and perform an evasive manoeuvre with its own power.
  BUT: If an obstacle is encountered, the AGV must stop and inform staff so that the obstacle is moved out of the way, because it should not be there. In an environment characterised by fixed machines and a network of drive tracks, it does not make sense to cheerfully start an evasive manoeuvre when an obstacle appears. Imagine the analogy in road traffic!
- The autonomous AGV is able to think up for itself the route on which it will drive.
  BUT: The AGV master control system knows all transport orders and the utilisation of the vehicle fleet. Of course, it issues the transport orders to the AGVs so that optimisation takes place at a central point and the predictability of the processes is maintained!
- The autonomous AGV is able to search for and execute transport orders by itself.
  BUT: Vehicles that autonomously search for transport orders are a horror scenario for orderly structures.

- The autonomous AGV does not need an AGV master control system.

  BUT: The central functions of the AGV master control system are transport order management, vehicle dispatching and transport order processing. Of course, one can imagine that a fleet of AGVs can manage itself without a master control system, but the functions mentioned must still be fulfilled, and in the absence of the AGV master control system, by the autonomous AGVs themselves. Powerful vehicle controllers and an enormous need for wireless communication are the consequence for the functioning of the whole fleet of vehicles and of course the assumption that all AGVs come to the same results in their considerations/calculations. So then we just have a shift of software services from a centralised structure to the decentralised one.

In the automotive industry, the uncertainty regarding the products to be built in the future is so great that they are developing ideas (of course, they also love to do research) of a *production of the future* that is not only supposed to be flexible, but hyper-flexible (choice of words analogous to "active children" and "hyperactive children"): Everything is supposed to be possible, there are no fixed machines or structures, it is supplied by drones, and everything is in a constant state of flux. In such an environment, autonomous functions of the AGVS seem to make sense, or let's say: such functions don't even bother there, because everything is a priori chaotic/agile.

This is the vision of the future of a factory without people, where robots do everything and people are no longer needed—science fiction. However, experience teaches us that in intralogistics, classic processes are still the order of the day, with strict adherence to performance requirements and strict economic specifications. There is no room for science fiction.

The AGV in intralogistics is not autonomous, but driverless, automatic, calculable, safe, reliable and economical. And it needs certain structures (routes, floor condition, safety distances, trained personnel, etc.) so that it can permanently be the backbone of efficient production logistics.

These remarks are sincerely meant and should help to ensure that the terms *autonomous* and *automatic* are used with care. Often enough, one encounters the term "autonomous", which is only used to make products and services appear "sexy". Unfortunately, the above remarks and the appropriate term "automatic" are not "sexy" at all.

4.4.2   Autonomous Interaction: Intelligent Action

Today's AGVs are controlled centrally. The AGV master control system takes over the central functions and instructs the vehicles. These are only supplied with the information they need to carry out their task. All coordination tasks lie with the master control system.

Fig. 4.14 First steps towards autonomous AGV. (Source: AGILOX, A-Vorchdorf)

The more intelligence we expect from the AGV system, the more intelligent the individual AGVs will have to be. If we want "seeing"[15] vehicles, they will have to obtain and evaluate all the necessary information themselves. To achieve this, there will be increased communication with peripheral equipment, but also among the AGVs. The vehicles will inform, warn and, if possible, help each other.

A concrete wish would be, for example, to be able to use an intelligent vehicle—let's think of a forklift AGV or a tugger AGV—as an initial single-vehicle system without much installation effort. If one then brings in a second or third AGV, this small group could work immediately without having to add an AGV master control system and higher-level functionalities. The vehicles could be so self-explanatory that they could be put into operation by the operator, without support from the AGV system supplier.

With the AGILOX (Fig. 4.14), the supplier meets the demand for intelligent behaviour. Some forward-looking features have been realised here; these include the fact that the AGVs have all logic on board. They navigate by means of contour navigation and fulfill all tasks without any master control system.

There has been research on the subject of swarm intelligence for some time. Researchers are investigating why flocks of birds fly straight ahead, although the individual birds move

---

[15] The desire for "seeing" AGVs does not a priori mean focusing on stereo cameras as central sensors. Which techniques will prevail is rather unimportant for the matter.

Schiller
intralogistics automation

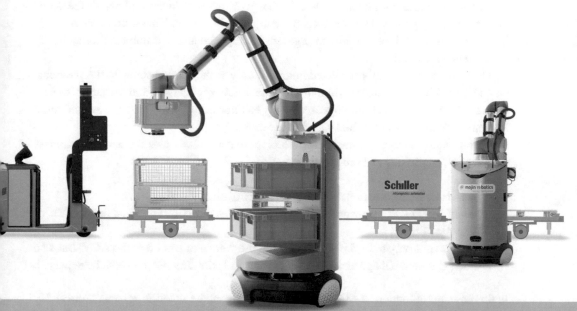

<9 mojin robotics

HILLER MOBILE ROBOTICS · autonomous modular intralogistics

Member of **ERIN**®

in a criss-cross pattern within the flock. Ant or bee colonies also act very effectively and successfully as a group, whereas a single animal is lost in the world. It seems foolhardy to try to transfer these considerations to an AGV with many vehicles—although the AGV manufacturers probably like the idea of many AGVs. Nevertheless, from these basic investigations come hints on how to organise intelligence and decision-making authority in the vehicle network.

The behaviour of AGVs predicted here will only make limited sense in the previous/ classical AGV applications, it will rather be about new applications and perhaps also new markets. The use of small intelligent vehicles has not been possible so far, which could ultimately be the reason for the lack of applications.

Autonomous functions in intelligent vehicles are particularly useful when the fuzziness of operating conditions and orders increases:

- If there is no defined path network, but a freely available action area.
- When it's no longer "Transport from A to B in a given time!", but "Find pallet XY!" or "Clean the floor!".
- If there are not only trained personnel within the area of operation of the AGV, but also visitors, children, old and sick people, i.e. careless actions by people are to be expected.

In this case, the autonomous AGV must act depending on the situation, it cannot (only) rely on the AGV master control system, but must determine itself, i.e. use its freedom of decision and action in the sense of safe operation. In such cases, however, the performance specifications are probably not in the foreground! Here is an abstract example: How long an autonomous cleaning robot needs to clean the arrivals hall of an airport cannot be predicted exactly because it depends on many factors, especially on the number of passengers staying in the area. And a concrete example: At the Leipzig plant, BMW uses a "SortBot", this robot sorts empty containers.[16]

We have limited ourselves here to the use cases in the internal environment of intralogistics. However, the primary applications for autonomous vehicles are more likely to be in public or military areas. And autonomous functions will certainly also find their way into intralogistics; we just want to point out once again that the demand for autonomy is not always and necessarily sensible, but should be well considered and justified.

---

[16]Marco Prüglmeier, project manager at car manufacturer BMW, in the forum "Transports and handling—autonomous all by themselves" on 14th March 2018 at LogiMAT in Stuttgart.

The Holistic AGVS Planning

5

First of all, it must be clear that the adjective *holistic* is used here deliberately and dialectically correctly. AGV planning is to be presented comprehensively here—as in VDI Guideline 2710,[1] which will be cited several times; in other words, all steps of planning are to be addressed, beginning with the first preliminary considerations and ending with the decommissioning of the old plant. Obviously, we will limit ourselves here to the AGV aspects—there are plenty of general presentations on the subject of project management.

In the previous chapters we got to know the AGV as an organisational tool that plays an integrative role in the intralogistics environment. There are many points of contact—i.e. technical or organisational interfaces—with secondary, subordinate or superordinate topics and trades, so that the planner quickly realises that he has to deal with much more than just the pure AGV technology.

Until today, we have been dealing almost exclusively with a project business. There have been and still are considerations to market AGVS, or at least certain AGV series, like a product. Until today, however, this is rather an exception or a market niche. There are hardly any two identical systems; experience shows that follow-up projects also always differ. So it is a tangible project with many facets when it comes to planning AGV systems. This is not intended to cause fear, but only to put the importance of AGVS planning in the right perspective.

5.1   The Importance of Planning in AGV Projects

AGV systems are of varying complexity. The range of realisations extends

---

[1] VDI 2710 "Holistic planning of automated guided vehicle systems (AGVS)".

© Springer Fachmedien Wiesbaden GmbH, part of Springer Nature 2023
G. Ullrich, T. Albrecht, *Automated Guided Vehicle Systems*,
https://doi.org/10.1007/978-3-658-35387-2_5

- from simple low-cost systems to high-end intralogistics solutions,
- from plants with only one to well over a hundred vehicles,
- from indoor to outdoor use,
- from functionally simple (transport from A to B) to complex systems that fulfil other demanding functions beyond the pure transport task,
- from point-to-point connections to intelligent taxi operations,
- from very small to very large vehicles (the load weights range from a few kilograms to more than 100 tons).

The user's task can range from simple point-to-point transport tasks with only one vehicle to the implementation of comprehensive logistical operating concepts in which the AGV is merely a component within a production, storage or distribution area. The degree of integration into the overall operation and the scope of the task inevitably determine the depth of planning. With growing complexity, in addition to the pure AGV functionality, there are increasingly questions about the appropriate modification of e.g. buildings, warehouses, assembly lines, production machines, etc.

Therefore, the logistical solution must be embedded in an overall concept, whereby various interfaces to secondary, superordinate and subordinate systems must be taken into account.

As with every project, besides the purely technical, process-related and infrastructural aspects, the economic aspects essentially determine the outer framework of the planning. Carefully prepared proofs of economic viability are required for the necessary investments. Alternative financing models must be examined.

In addition, the AGV planner is often occupied with secondary issues that are rather rare in the daily planning business: For example, the project has an external impact (within the company group or even on the industrial sector) or consequences for the employees (social or labour law), so that it is intensively observed by third parties.

Therefore, it is important for the planner to take the holistic aspect of his task into account from the very first moment in order to make the AGV project successful. This is because all relevant aspects must be taken into account with the necessary conscientiousness and brought into a causal relationship.

Ultimately, AGVS planning is quite complex and demanding. Not every potential user has the necessary know-how or the required capacities/resources within their own ranks, which makes external planning support sensible. Too often, the dilemma between planning complexity and lack of resources leads to the abandonment of what would be a sensible AGV solution in favour of a simple transport solution (e.g. with conventional manual industrial trucks), wasting a great opportunity to increase efficiency every time.

5.1.1   Resource-Determining Criteria

How many resources does an AGVS project require? Certainly, optimal AGVS planning depends on many factors, these factors essentially include:

Step-by-Step Automation

We put theory into practice

Toyota Autopilots take the flow of goods to a new dimension of productivity. The AGV's are in operation 24 hours a day, seven days a week and perform repetitive tasks – saving valuable resources such as time, costs and materials.

TOYOTA

- Scope of the overall project:

    Is it a pure AGVS planning (stand-alone project) or is the AGVS only one trade in the overall network of a larger material flow project? Will the AGVS be tendered separately or will it be negotiated and awarded as a package with other systems?
- New or modernisation planning:

    The terms *greenfield project* and *brownfield project* are also frequently used. A greenfield project means the new construction of a building (on a greenfield site, so to speak), while a brownfield project means the construction of a new system on or in an existing structure from previous use, often combined with installation and commissioning during ongoing operation. This is certainly a decisive factor in how much the adjacent systems can be adapted to the AGVS, or to what extent the AGVS has to integrate into existing structures.
- Budget:

    The size of the project budget will also influence the planning. Ultimately, the budget is also a measure of the complexity of the project.
- Impacts in intralogistics:

    How big are the effects of the use of AGVs on the operational process and on other trades? The more networked the systems are, the higher the demands on holistic planning.
- Importance of the use of AGVS use for the logistics process:

    How great will the dependence on the functioning AGVS solution be? Will an AGVS failure lead to a standstill in production or delivery? Does the availability of the AGVS directly determine the performance of the entire production?
- Existing experience with AGVS:

    Generally speaking, companies that already use AGVs find it easier to plan new systems. Often this advantage is related to individuals.

Resources for planning are allocated according to these criteria. Because without the appropriate resources, careful AGVS planning is not possible.

In this context, we should address the frequently asked question of how long an AGVS project takes. Of course, due to the "multi-dimensional span" of the project, only a rough indication can be given here. In times of high demand, a period of 10–12 months is estimated for the pure manufacture, delivery and commissioning of an AGVS. If you add the tendering period, i.e. from the first pages of the specifications to the awarding of the contract, of about 6 months, the actual project duration is easily one and a half years. To this must be added the time for "preliminary considerations", i.e. finding the concept. As a rough rule of thumb, 2 years can be assumed.

5.1.2   Organisation of the Project Team

At the same time, and due to sector and/or company specifics, it is also decided who will ultimately carry out and be responsible for the individual planning tasks:

- Independent planning without/with external support

  In-house planning is carried out by the company's own staff or by external staff commissioned by the company. It should be noted that the employees must have detailed knowledge of the existing procedures and internal processes, otherwise time-consuming training will be necessary. Of course, the planning work means that they are not available to the rest of the operational business, or only to a limited extent. In the case of "larger" projects that are being implemented for the first time, it is advisable to seek external support.

- Supplier planning

  Planning can also be done directly by the AGVS manufacturer, although this approach carries the risk that only a sub-optimal solution will be found, as the manufacturer will often only draw on its own delivery programme for planning. In addition, it is necessary to decide on a supplier at a very early stage. With such planning, the fulfilment of the specified performance data and characteristics is the sole responsibility of the supplier. In general, such an approach is based on a so-called *functional tender*, which describes a logistical task but leaves the solution open except for possible main components or the transport principle. Supplier planning makes sense if this supplier has already delivered solutions for similar tasks in his own company and has thus strongly recommended himself. His experience advantage over other suppliers additionally speaks in favour of such a solution.

- General contractor project

  A different distribution of responsibilities can be achieved by awarding the contract to a general contractor. In this approach, which is suitable for more complex projects, all subsections of the overall project (e.g. construction measures, steel construction, mechanical engineering and plants, etc.) are handed over to a partner responsible for the project. This partner can in turn assign individual trades to subcontractors. The characteristic feature, however, is that there is only one contractual partner vis-à-vis the customer, who is responsible for the performance of the entire plant. Holistic planning is ensured by the general contractor, who can, however, call in external specialist planners for this purpose.

- Mixed forms of the above.

5.2   Planning Steps

We structure AGVS planning according to VDI 2710. Firstly, because it makes a priori sense to accept the content of a valid technical set of rules, and secondly, because the author played a major role in the creation of said guideline. Therefore, it would also be untrustworthy if these pages deviated from the content of the guideline, since the creation was only 18 months apart.

Table 5.1 The planning phases of an AGV system according to VDI 2710

Pos.	Designation	Result
1	System finding	System decision in favour of an AGV system has been made, proof of economic viability has been provided
2	System design	Specifications available
3	Procurement	AGV system is installed and ready for operation
4	Operational planning	AGVs are operated reliably
5	Change planning	AGV system is changed
6	Decommissioning	AGV system is disposed of

VDI 2710 is aimed at interested parties, planners, operators and manufacturers, can serve as a planning basis for AGVS projects and is intended to promote mutual understanding.

Like any project, the realisation of an AGV system requires a step-by-step approach that can be checked and coordinated at any time (Table 5.1).

At the beginning, there are considerations about the type and scope of the project, the intended implementation model, i.e. the question of complete redesign or integration and expansion of existing structures, and the financial and economic framework conditions.

In this way, a "script", a scenario for the planned project is created and described, which, in addition to the original task, must above all consider all networking points in the operation. Since smooth operation cannot always be assumed, especially during system ramp-up, it makes sense to also consider the influences of failures and to work out alternative courses of action as a precaution.

If the project has been described in sufficient detail by the idea providers in advance, the next step is to organise the project properly.

The structure of a project team is oriented towards the intended planning variant, i.e. whether the project is essentially the responsibility of the company itself, a supplier or a general contractor. In connection with the question of whether experts from outside should be brought in, this determines the degree of involvement of the company's own staff and, above all, the selection of a competent group of employees.

The planning phases do build on each other logically. Only in rare cases, however, will it be possible to work through the individual steps successively one after the other. Corrections to previous planning phases will often be necessary. Such feedbacks can be, for example:

- During the detailed technical planning, the planner notices that technical details call the entire concept into question; example: explosion protection manners required for the AGV (Sects. 5.2.2 → 5.2.1)
- The profitability considerations are not accepted by the management; the concept must be changed (Sects. 5.2.2 → 5.2.1)
- The analysis of the market does not conform to the specifications (Sects. 5.2.3 → 5.2.2)

- The offers do not correspond to the adopted budget plans (Sects. 5.2.3 → 5.2.2)
- Running costs are too high due to special technical details (Sects. 5.2.4 → 5.2.2)
- The calculated replacement and expansion investments go beyond any scope (Sects. 5.2.5 → 5.2.1 or 5.2.2)

5.2.1 System Identification

The aim of this phase is to make a decision in favour of the AGV system. If the result of the concept finding is against an AGVS, but in favour of an alternative conveyor technology, the AGVS planning already ends here. Criteria should therefore be mentioned here, the examination of which leads to a system decision "for or against" an AGVS.

5.2.1.1 As-Is Analysis

The primary objective of an as-is analysis is to generate initial data for planning and to identify weak points. The information gained from the as-is analysis is the starting point for the subsequent planning and implementation steps.

When planning material flow systems, the highest demands are placed on the as-is analysis for the future-oriented design of systems and organisational forms:

- Delimitation of boundary conditions,
- recognise existing potentials,
- plausibility checks,
- transparent presentation of work processes,
- provision of key figures for the optimisation of material, information, energy and personnel flows,
- compiling information with maximum information content,
- ensuring the greatest possible planning security.

The planning bases obtained within the scope of the as-is analysis provide the planner with a pool of meaningful key figures. The prerequisite is that the scope of the as-is analysis is clearly defined and that suitable methods and tools are used for data collection.

5.2.1.2 Needs Analysis and Concept Identification

The need for a new solution results from the requirements of the future, the experience gained in the past and the current analysis of the current situation.

- Past: Accumulated experience
  This point is omitted in new planning. Otherwise, extensive experience could be gained in the operation of the existing facility, which does not have to be an AGV solution. This is both quantifiable experience that can be incorporated into the as-is analysis, and qualitative experience that has to do with the acceptance and up-to-dateness of the technology used.

- Present: Actual analysis

  The as-is analysis is the official summary of the found situation and thus an important data base for the documentation of the project.

- Future: Planning data for the company's development

  The first step is to agree on reliable planning data as planning specifications. This must be done together with all relevant departments in the company. Since it is often the case that technical planning fails to meet the requirements because the planning data changes in the course of the project, it is important to record this data at the beginning and during each project.

Ultimately, this planning step has a decisive character. Different solution concepts are played through and compared with each other; typical questions are:

- Level of automation?
- Clocked lines, island concept or flow lines?
- Arrangement of the material, logistical concept?
- Technical equipment of the production facilities, loading aids?

A material flow simulation can be helpful here and can provide answers.

If the planning envisages a significant increase in the degree of automation with the project, then the needs analysis should definitely have visionary aspects. Here, for example, general goals can be formulated that cannot be easily evaluated in monetary terms:

- The forklift-free factory (no damage from conventional man-operated forklifts),
- more cleanliness and order in production,
- process reliability through controlled processes,
- positive external impact on customers and partners,
- adding value to the site through high quality, performance and reliability.

5.2.1.3 Framework Data

The VDI's AGVS checklist[2] provides support in collecting the framework data. The completed checklist is used for the complete recording of all relevant data required for the preparation of a specification. It is divided into the following sections

- General description of the task,
- conveyed goods, conveyance aids and loading unit,
- load provision,
- layout,

---

[2] VDI 2710 Sheet 2 "AGVS checklist—planning aid for operators and manufacturers of automated guided vehicles systems (AGVS)".

- material flow with source/sink matrix,
- energy system,
- control with analysis of information flows,
- surroundings,
- special requirements.

The AGVS checklist is a planning aid for operators and manufacturers of automated guided vehicles. It contains a catalogue of all relevant planning data to describe the planning task and the planning environment. This checklist includes the information that the later operator of the AGV must compile so that, on the one hand, he gets clarity about the planned AGV use and its organisation and, on the other hand, he can provide the manufacturer with detailed information about the requirement profile for the conveyor system.

Note: Although this guideline is written for AGV systems, it can probably also be used for other technologies in principle, but must then be adapted in detail.

5.2.1.4 System Selection

Together with the framework data and the needs analysis, the basic data for a system decision are now available. These help with the necessary fundamental considerations regarding the process to be planned. Examples for these basic considerations can be:

- Targeted level of automation,
- logistics concept,
- selection of a loading aid.

Ultimately, a decision has to be made on the support system to be used. The choice is made on the basis of technical as well as economic considerations.

Technical System Selection

Another VDI guideline helps in deciding on the appropriate conveyor system on a technical basis.[3] It gives the logistics planner assistance in the technical selection of a transport system. By excluding those transportation technologies that are disqualifying for his task, it leads him to a list of transport systems that are suitable for his task. The guideline thus ensures that all common transport systems are considered in the selection process. By describing the main features, characteristics and suitability of the conveying systems, the planner receives important information for proper selection. This enables the planner to assess the common conveyor systems.

The guideline provides an EXCEL[4]—based tool that implements the tables from the guideline and thus facilitates the application of the guideline. The definitions and

---

[3] VDI 2710 Sheet 1 "Holistic Planning of Automated Guided Vehicle Systems (AGVS)—Decision criteria for the selection of a transportation system".

[4] EXCEL is a product name and trademark of Microsoft.

explanations of the chaining principle and the relevant framework data are set out in the guideline. This Excel tool is amazingly easy to use and is also not intended for the confirmed AGV expert, but rather for the topic novice and especially suitable as a training tool.

Economic System Selection

In order to also support the economic side of system selection with a VDI guideline, VDI 2710 Sheet 4[5] is recommended: It is primarily aimed at investment planners who are faced with the task of holistically evaluating the economic viability of automated guided vehicles. The extended economic efficiency analysis described consists of the parts economic efficiency calculation, utility value analysis and overall evaluation. It is suitable for automated guided vehicles and other capital goods that are characterised by high acquisition costs, a long useful life, and many properties that are difficult or impossible to quantify in monetary terms.

An overview names and explains different types of costs for a detailed economic efficiency calculation, including aids for determining guideline values for the types of costs. Dynamic methods of economic efficiency analysis are described by means of an example. The utility analysis integrates criteria into the extended economic efficiency analysis that cannot be quantified in monetary terms or only with unjustifiably high effort. They are listed in a structured manner in the guideline. Finally, it is shown how an overall assessment can be made on the basis of the results of the economic efficiency calculation and the utility value analysis.

Since the extended economic efficiency analysis requires a lot of computing effort, especially with dynamic methods, an EXCEL-based tool is included that does the calculations after the data has been entered. Profitability analyses of up to three transportation technology alternatives are quantified and graphically displayed. In addition to the investments, a distinction is made between the direct and indirect costs as well as the additional benefits (see Tables 5.2, 5.3, 5.4, and 5.5).

In addition to the economic efficiency calculation methods used in VDI 2710 Sheet 4, the following methods are frequently used:

ROI Consideration

In financial terms, the term *return on investment* (ROI) means the ratio of the profit achieved from an investment to the capital invested. This ratio can be related to a defined period of time as well as accumulated over the entire life cycle of the investment. For example, the costs of decommissioning and disposing of plants, which subsequently reduce profits, must be taken into account. When naming the ROI, it should therefore always be

---

[5] VDI 2710 Sheet 4 "Analysis of The Economic Efficiency of Automated Guided Vehicle Systems (AGVS)".

Table 5.2 Investments of an AGVS calculation

Position	Description
AGVS	Vehicles, control system, ground system, project-related service
System peripherals	Load transfer stations and buffers, if credited to the AGVS and not to the stationary conveyor technology
Structural measures	Floor renovation, protective devices, adaptation of fire protection gates, bridges and ramps
Integration into existing structures	Interfaces to superordinate, subordinate or secondary controls, integration of automatic scales, scanners etc.

Table 5.3 Direct costs in an AGVS calculation

Position	Description
Maintenance	The even and gentle driving style minimises wear and tear on tyres, batteries, drives, etc.
Energy	Essentially the charging current for the traction batteries
Personnel of the plant operation	Only related to the transport system; control desk personnel only proportionally!
Taxes and insurances	
Transport damage to the product	Automatic transport minimises transport damage. Material, extra work and rework, but also customer complaints have to be taken into account
Transport damage to operational equipment	Such as loading aids, columns, walls, racks, shelves, gates

Table 5.4 Indirect costs in an AGVS calculation

Position	Description
Personnel costs in related areas	If necessary, forklift drivers, personnel for pallet provision and for fine distribution
Stocks	By improving the flow of information and high availability, stock levels can be reduced
Material stocks in production	
Lead time	The order duration is reduced and the order density is increased—thus increasing the efficiency of production

stated with which framework conditions (time span, consideration of which costs, etc.) it was determined. There are often internal company guidelines for this.

Colloquially, *return on investment* is also used in the sense of the payback period, i.e. the time between the investment being made and the point in time at which the income generated by it has accumulated to the level of the investment sum.

Table 5.5 Additional benefits of an AGV System

Position	Description
Flexibility and adaptability	Flexible use of space, adaptation to transport fluctuations, material flow and layout changes
AGVS as an organisational tool	The AGVS master control system ensures an optimal flow of materials and information and thus more transparency
Minimisation of incorrect deliveries	Automation ensures absolutely reliable transports and a high level of process reliability
Safety	The AGVs operate safely and accident-free
Order and cleanliness	Stress is reduced and a pleasant ambient atmosphere is created
Availability and continuity	The AGVS works unspectacularly, without interruption, without any hustle and bustle
Ecological benefits	Low sound level, no emissions, low energy consumption
Ideal advantages	Showcase production, image impact internally and externally, technological edge

Total Cost of Ownership (TCO)

When assessing different solutions of a technical problem, it is becoming increasingly necessary to carry out an extended evaluation of the costs. Whereas, for example, in the case of a procurement measure with the same technical performance of the alternatives, only the different investment amounts were previously compared, the monetary effects indirectly associated with the measure must increasingly also be taken into account in order to find an optimal solution with regard to all the expenditures induced by the decision.

The "Total Cost of Ownership—TCO" method expands the perspective beyond the pure investment and thus makes it possible to define and quantify downstream costs, as well as to provide starting points for cost optimisation and contract negotiations.

When planning an AGV system, the result of the profitability analysis of the system plays a decisive role due to the relatively high investments. The arguments for the use of an AGVS result on the one hand from the low operating costs and on the other hand from a considerable additional benefit. In many cases, this additional benefit, e.g. the high reliability of the delivery of the transported goods, cannot be quantified in monetary terms, or only with a great deal of effort. However, as it is a key factor in the economic efficiency of the transport system, it must not be neglected in an economic efficiency analysis.

Important: At the end of the system design, the results of the economic efficiency calculation must be checked again, because the level of detail is then significantly greater than before.

5.2.2   System Design

The aim of this phase is to plan the project in detail and to draw up meaningful specifications, as well as to adjust the economic viability of this detailed planning within the company and thus to check it again. For complex systems, a simulation is helpful, the specifics of which will be briefly discussed here.

System planning includes an essential aspect that is unfortunately often forgotten: Timely coordination with the relevant employers' liability insurance association and the labour inspectorate. Those responsible in occupational health and safety (both within and outside the company) must be included in the planning in good time.

5.2.2.1 The Simulation

Simulation is the reproduction of a system with its dynamic processes in a model that can be experimented with in order to arrive at findings that can be transferred to reality.[6] In contrast to this, static observations use average values, e.g. of throughput times, workloads, speeds, etc. The dynamic simulations use more complex distributions with time-based fluctuations and interactions. In addition, random number generators are used to replicate sudden and unexpected events.

A flow simulation models the entire process chain of the material flow. It takes hold and delivers useful results for the following three areas:

- Production planning: layout planning, material flow and production control, synchronizing, working time models and resource planning
- Factory logistics planning: supply concept, layout and resource planning
- Supply chain planning: supplier selection, JIT/JIS concept, supply chains (Fig. 5.1)

Related to the AGV, the simulation can provide the following results:

- Review of the logistical concept (performance and capacity utilisation),
- optimisation of disposition strategies when varying the layout or the control system,
- required number of vehicles,
- dimensioning of load transfer and storage places.

The advantages of a simulation in the run-up to realisation are the saving of time and costs, the increase in planning reliability and thus the minimisation of entrepreneurial risk. In addition, the following side effects usually occur:

- Creation of a common database for all project participants,
- vivid and verifiable basis for discussion during planning,

---

[6] VDI 3633: Simulation of Logistics, Material Flow and Production Systems, March 2000 and VDI 2710 Sheet 3 Fields of Application of Simulation For Automated Guided Vehicle Systems (AGVS).

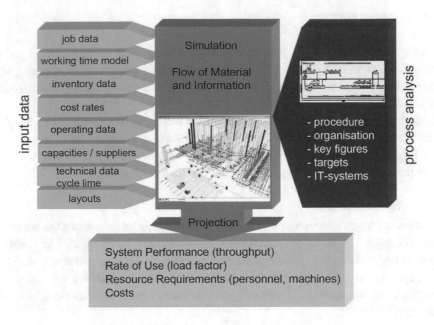

Fig. 5.1 Simulation of goods transport systems. (Source: SimPlan)

• 3D animations and full-scale overview of the project.

Every simulation has fundamental limitations and weaknesses. Its results are based on the input data. If these are faulty or incomplete, incorrect simulation results are foreseeable or unavoidable ("garbage in—garbage out"). Not all eventualities and special cases can always be taken into account, which is why we should warn against excessive expectations at this point.

5.2.2.2 Technical and Organisational Delimitation of the AGV System

The so-called worst-case scenario is an effective means of checking planning assumptions. For a defined period of time, the peak performance of production machines and simultaneous disruptions in adjacent trades are assumed. On the one hand, this can lead to very high target performances of the AGV system or, on the other hand, to organisational solutions. These exceptional cases can be countered with an increased number of AGVs or, for example, with a calculated number of buffer positions.

It is important to think through all eventualities, especially if the dependence of the overall logistics process on the AGVS is high. This is the right time to think about the possible failure of system components. This concerns the AGVS, but also its periphery.

It may sound banal, but it's true: An AGVS does not fail. The power supply for certain areas of the factory fails, or a computer, or a computer network (LAN or WLAN), or an AGV. But not an AGVS.

It is therefore necessary to quantify the individual probabilities of default and assess the consequences. If the consequences are high, a plan B is needed. That means, for example:

- The failure of a PC (primary system) on which the AGV master control system (or parts of it) runs must be safeguarded by a "warm standby" system. This means that there is a second, secondary computer that runs in the background and quickly takes over the tasks of the failed primary computer in case of failure.
- The networks must be built according to the rules of modern IT systems, which is not an explicit AGVS topic.
- The failure of an AGV must be planned for. In systems with several vehicles, the failure of a single AGV must not lead to the necessary performance of the AGVS no longer being possible. Or an emergency strategy—possibly with man-operated forklift support—must be put in place.

In this context, the operator, i.e. the client,[7] should check whether the hardware of the host computer level as well as the networks (LAN and WiFi) should not be provided by the client, because then the system and data security is integrated into the IT world of the company and does not mean an insecure parallel world.

In addition, the peripheral interfaces of the AGV system to the outside world must be considered. These include the active load transfer stations, lifts, automatic doors and gates, but also all data interfaces, e.g. to systems from which the AGVS receives its transport orders. A fallback strategy must be created for all these interfaces at this stage.

5.2.2.3 Detailed Technical Planning

Detailed technical planning now means the technical assessment of the technology to be used. A lot of information on this can be found in Chap. 2. In the following, we will only refer to a few specific points of detailed technical planning.

Layout Planning

Here, the initially rough layout (see framework data) must be concretised. For this purpose, the material flow relationships, the transport volume, the transport times and the conditions in the production environment must be taken into account. A layout should in any case be the basis of the tender, or is to be prepared by the supplier in the course of the tender preparation. It is advisable to define the directions of travel on the individual route sections.

It is common to assume right-hand traffic, as it is easier for other road users to adapt to driving movements. A good layout is characterised by simplicity. One-way traffic and organisational separation of industrial trucks and visitor flows are just as much a part of this as the fewest possible intersections.

Calculation of the Number of Vehicles

For simple applications, the required number of vehicles can be determined by means of a source/sink transport matrix, as already established in the framework data. VDI Guideline 2710-2 describes a simple method for calculating the number of vehicles based on a

---

[7] Client = principal, counterparty to the contractor = supplier.

transport profile. In addition to the source/sink matrix, the layout and the operating times are required.

For more complex applications, a simulation makes sense. There are various systems and providers on the market for this purpose. To determine the number of vehicles, the prerequisites must be defined at the beginning of the simulation: Number of working shifts, operating times, break model, handling times of the transported goods, possible conveying speeds, degree of blockage, energy concept, operating concept, repair strategies, technical availability, layout should be available, emergency operating concept.

A too high number of AGVs is a price driver and sometimes even means the end of the AGVS project. Therefore, it is particularly important to check whether

- Power peaks are absorbed by buffer spaces or an intelligent control system and/or
- the loss of individual vehicles can be compensated for at short notice by other funding.

Depending on the operational concept, additional vehicles may have to be kept in reserve (repair and maintenance).

An interesting question in this context is who specifies or is responsible for the number of AGVs. Here, the user, the planner or the AGVS supplier come into question. It is certainly helpful for the comparability of offers to request a fixed number of AGVs. In this case, however, the responsibility for the performance of the AGVS lies with the user or planner. If performance on the basis of the transport matrix is a significant part of the specifications, the supplier of the AGVS bears the responsibility. In this case, however, the incoming offers may differ in the number of AGVs. Wherever the responsibility may lie in the respective project—it is important that all parties involved are aware of it!

AGV-Compatible Building Planning

There are two possible variants for building planning that is suitable for AGVS: Planning into a new building (greenfield) or integration into an existing building (brownfield).

In greenfield planning, the concerns of the AGV technology can be optimally taken into account by controlling the requirements at an early stage. The design of the ground plate, floor surface, building supports, building heights, façade, energy requirements, media, etc. can be tailored to the AGVS requirements. Brownfield planning, on the other hand, can only react to the existing conditions.

A floor suitable for AGVs is essential for the safe and trouble-free operation of AGVs. Its characteristics are described in Sect. 2.4.1 and must be observed when manufacturing new floors. If existing floors do not meet all these requirements, timely consultation with the AGV manufacturer is strongly recommended.

The following standards and guidelines are authoritative for the planning of traffic routes:

- Workplace Ordinance,
- Workplace Guideline ASR 17/1,2,
- DIN 18225—Traffic routes in buildings.

Table 5.6 Detailed technical planning of peripheral facilities

Position	Description
Requirements for the operating environment	– Ambient conditions – Floor condition (clarify at an early stage!) – Traffic routes
Stationary facilities for navigation	– Installation of guidelines on the floor – Installation of guidelines and primary conductors in the ground – Installation of spot floor markings – Reflector installation
Stationary equipment for load handling	– Load transfer stations – Safety relevant aspects
Communication systems	– Data radio, narrowband/broadband radio (mostly WiFi) – infrared communication – Other
Stationary equipment for electrical power supply	– Battery maintenance and care – Stationary power supply equipment
Stationary safety devices	– See Table 2.8
Peripheral facilities, building facilities	– Doors and gateways – Fire section gates/fire shutter doors – Lifts – Lifter/Vertical conveyor – Crane systems – Track wagons/underfloor systems

If the prescribed safety distances and safety measures cannot be observed, coordination with the responsible employers' liability insurance association or the State Agency for Occupational Health/trade supervisory authority is required. Separate areas or rooms may be required for the operation of an AGV system. These may be battery charging stations, service and maintenance areas, for example. Further details can be found in VDI Guideline 2510-1.[8]

Infrastructure and Peripheral Facilities

More details on this complex topic can be found in Sect. 2.4. Table 5.6 indicates where requirements are placed on the adjacent trades from the AGVS' point of view.

Especially the control integration of the AGV system into the periphery can lead to unexpected costs. Often, the data connection to fire shutter doors or lifts is not standardised and means greater effort as well as additional authority approvals (Fig. 5.2).

---

[8]VDI 2510 Sheet 1 "Infrastructure And Peripheral Equipment For Automated Guided Vehicle Systems (AGVS)".

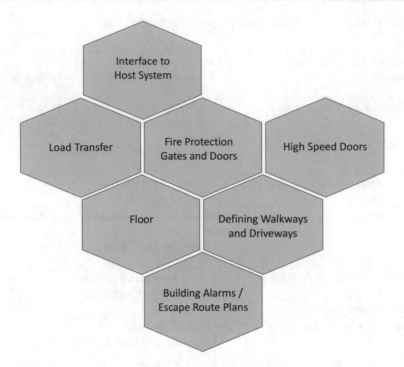

Fig. 5.2 Important "interfaces" as pieces of the mosaic of successful planning. [Source: Siemens (H. Kohl: *AGVS planning from the operator's point of view*. Lecture on the occasion of the LogiMAT, Messe Stuttgart, on 05.03.09, Siemens AG, Industry Sector, Frankfurt am Main)]

Material Flow Control
This refers to the interaction of the AGV master control system with the higher level, i.e. with the computer world from which the AGVS receives the transport orders. Material flow control refers to the recording and control of transport processes and sequences with the aim of managing the transport volume with intelligent rules but as few AGVs as possible. It is an important factor because it has a direct influence on the number of vehicles. As the degree of automation increases, so do the demands on material flow control. Reference is made to Sect. 2.2 and VDI Guideline 4451 sheet 7 "AGVS Guidance Control System".

5.2.2.4 Specifications
The specifications describe the totality of the requirements for the deliveries and services of a contractor in a result-oriented manner. In principle, the client should write down the specifications. It then serves as the basis for the invitation to tender. Sometimes the specifications are also referred to as list of services (LOS).

It is nevertheless widespread for the potential contractor to draw up the specifications himself in consultation with the client. This has the great advantage for the contractor of being able to define the service to be provided by him. For the client, this results in the risk that the contractually agreed service does not exactly meet his needs.

The content of the specifications should include the following points:

- Specification of the product to be created,
- requirements for the product in its subsequent use,
- framework conditions for product and service provision,
- contractual conditions,
- requirements for the contractor,
- requirements for the contractor's project management,
- acceptance procedures.

In a formally correct procedure, the contractor, after receiving the specifications, translates the results to be delivered (loads) into required activities (duties) and later draws up the so-called final design specifications. The tender documents describe the scope of services to be delivered. The documents should reflect the commercial, organisational and technical requirements of the project.

- The commercial part regulates: General Terms and Conditions (GTC), basics for preparing a quotation, form of the quotation, pricing, payment terms, invoicing, measurement, cost tracking, contractual terms, ownership and usage rights, claims for defects, liability, gross scope of delivery, etc.
- The organisational part regulates: Project sequence, project implementation, project start, approval phases, assembly, electrical installation, commissioning, site installation, technical release, performance test, availability test, transfer of risk, trial operation, acceptances, safety, deadlines, etc.
- The technical part regulates: project-specific requirements such as operating times, cycle times, technical design, on-site services, project planning, documentation, detailed scope of delivery, mechanical-electrical delimitations, interface definitions as well as legal regulations to be complied with, such as standards, directives, (works) regulations.

Furthermore, reference is made to the VDI Guidelines 2519 Sheet 1 "Procedure For The preparation of Requirement Specifications" and VDI 2519 Sheet 2 "Requirement Specifications For The Use of Conveyor and Storage Systems". See also Sect. 5.2.3.4 "Requirements Specification".

5.2.2.5 Final Economic Efficiency Assessment

In addition to the specifications, the profitability calculations are part of the project documentation at this point. Within the company, there are adapted calculation formulas and considerations that must be applied to the project. As a result, specific key figures are calculated that must comply with certain predefined limit values. As far as possible, these methods have already been used in the planning step "system selection", but must be finalised here according to the detailed planning.

For further information, please refer to the section "Economic system selection" in Sect. 5.2.1.4.

5.2.3 Procurement

Within this planning phase, the AGV system must be tendered and procured, then it is installed and commissioned by the contracted supplier. At the end, there is the acceptance, with which the installed system is handed over to the operator.

5.2.3.1 Analysis of the Supplier Market

It depends on this planning step whether the right providers are really approached. Usually it is not an explicit planning process, but a process that prepares or accompanies the project. Trade journals are studied, large technology trade fairs (such as the Hanover Trade Fair[9] or LogiMAT[10] in Stuttgart) and conferences/congresses (AGV symposium[11] in Dortmund, Fraunhofer IPA Technology Forum in Stuttgart or VDI Material Flow Congress[12] in Garching) are visited or researched on the Internet. On the pages of the Forum-FTS[13] one can find the members of the European AGV community.

5.2.3.2 Tender

The specifications described above form the basis of the tender documents. The more carefully and completely the specifications are prepared, the easier it will be to evaluate the bids and award the contract.

It seems helpful to have a predefined price sheet that records the project components, structures them and queries them individually with price items. A rough subdivision for an AGV would be, for example:

- Vehicles incl. energy storage,
- floor system or periphery,
- AGVS master control system,
- project-related services.

The market analysis should have limited the number of providers requested. Typically, three to five independent offers are sought. In this context, reference visits are advisable. The three to five shortlisted suppliers are asked to organise a reference visit to one of their

---

[9] www.hannovermesse.de/industrial_automation

[10] www.logimat-messe.de

[11] www.fts-fachtagung.org

[12] www.materialflusskongress.de

[13] www.forum-fts.com

Table 5.7 Items of a bid evaluation

Position	Description
Technical part	– Total solution (weighted) – System components – General requirements for the fulfilment of the contract – Technology of the vehicles (AGV) – Transfer stations/parking spaces – Battery charging stations – Functionality of the AGVS master control system – Interfaces to higher-level systems, to the infrastructure and to future systems – Service – Options – Project Management – Schedule
Commercial part	– Investment – Total Cost of Ownership and/or ROI calculations – Energy costs – Spare parts prices/price fixing – Terms of payment, sureties, guarantees – Warranty period and conditions – Assumption of operational safety risk (emergency strategies)
Soft skills	– Type of tender response – Reference visits – Supplier assessment – Gentle load transport – Operating concept – Appearance of the product – Innovation

customers, where a similar or comparable system is installed if possible. These discussions with the operators always bring an information gain for the future AGV user.

5.2.3.3 Bid Evaluation and Contracting

The evaluation of offers and the contracting is described in Table 5.7. It is essentially based on the circumstances of the client. It serves to make the bids comparable and to obtain a uniform evaluation. Both the technical and the commercial scope of the offer is evaluated.

The process of contracting varies from sector to sector and from company to company. In industry, contracting negotiations are common, whereas in the public sector, submissions are mandatory. Bidder interviews are also increasingly used. Alternatively, "online bidding events" take place, often on supplier internet platforms.

After the contract has been awarded, the order is formulated contractually and accepted by the contractor with an order confirmation.

5.2.3.4 Specifications

In the VDI Guidelines VDI 2519—Sheet 1 "Procedure For The Preparation of Requirement Specifications" and Sheet 2 "Requirement Specifications For The Use of Conveyor And Storage Systems", these terms are defined and further instructions are given.

In the specifications (see also Sect. 5.2.2.4 "Specifications"), all requirements of the client with regard to the scope of delivery and performance are compiled. The requirements from the user's point of view, including all boundary conditions, are described here. In the specifications, the questions WHAT and WHEREFORE are answered, in the specifications the answers to the questions HOW and WHEREBY are given.

The specifications describe how the requirements of the specifications are to be fulfilled. The contractor lays down the client's specifications in detail and implementation requirements. While the requirements specification contains the specification of the requested AGVS as its core component, the functional specification describes how the contractor wants to/will provide the service. Thus, the work breakdown structure with the work packages is the minimum component of the functional specification. This includes the schedule and resource plans. For time-critical projects, the schedule becomes a binding part of the contract (contract schedule).

It is advisable to separate the requirements specification at least into a legal/organisational and a technical/professional part. The requirements specification and the specifications should be part of the contract between the client and the contractor.

In large projects with many partners, the specifications and requirements must be signed off by all partners. For critical projects, it is advisable to deposit the documents with a notary in order to be able to clarify subsequent claims beyond doubt.

In many cases, the creation of the functional specification is already seen as the first task of the realisation phase. The problem here is that a project "lives", i.e. theoretically the contents of the functional specification would have to be constantly adapted. A clear line must be drawn here, which should not be longer than 6 weeks after the start of the project.

5.2.3.5 Realisation

The project shall start after written assignment by the client and confirmation of the order by the contractor. At the beginning of the project, a kick-off meeting shall be held. The project managers and the project organisation of the contractor and the client shall be introduced at this meeting. The organisational procedure of the project—regular deadlines, communication channels, system of meeting minutes—shall be agreed between the contractor and the client. The agreed overall schedule shall be discussed and approved.

The procedure for technical clarification with approval documentation, sampling, increase/decrease procedure, change service or version management (drawings), creation of functional specifications, CE documentation,[14] hazard analyses, layout definition, integration into the production environment must be defined.

---

[14] CE = Conformité Européenne, English: in accordance with EU directives, within the scope of CE conformity in accordance with the Machinery Directive.

In the case of complex systems and tasks, it is advisable to plan and set up sample superstructures and test facilities. Within the framework of a FAT (Factory Acceptance Test), the preliminary acceptance of vehicles or partial trades/sub-works should take place at the contractor's premises, and the final SAT (Site Acceptance Test) at the client's premises.

The construction site phase for fixed technical systems is divided into mechanical assembly, electrical installation, commissioning and trial operation. When installing an AGV system, depending on the technology used, it can be much easier if, for example, complete and ready-to-operate vehicles are delivered by the supplier, so that the mechanical assembly is reduced to mere "unpacking".

Commissioning

Commissioning takes place after completion of the mechanical assembly and electrical installation. At this time, the contractor should request the preparation of defect lists from the client. During commissioning, the following measures, among others, are carried out: Functional tests (such as I/O tests), installation of software, step-by-step software parameterisation, commissioning of the individual components as well as the entire plant, plant briefings, safety-related initial commissioning, safety briefing of all parties involved.

At the end of commissioning, a functional test of the installation should be carried out under production conditions. At this point, the CE documentation should be available, including the declaration of conformity, which is in any case an important part of the documentation.

This functional test is then followed by trial operation. During trial operation, the AGVS is used by the contractor for the first time under the conditions of later production using original material to be conveyed. This is followed by the start of production.

Permits

Approval planning must always begin much earlier. The corresponding requirements shall already be part of the specifications and then later of the functional specifications. However, many required approvals are only granted after commissioning has taken place, so the planning step is described at this point. The general conditions for AGVs with regard to safety are determined by the following laws, regulations, etc:

- Machinery Directive, DIN standards, VDI guidelines, VDE regulations,
- accident prevention regulations,
- workplace regulations,
- guidelines of the professional associations,
- In Germany an approval by the VDS (Verband der Sachversicherer) is necessary.

Compliance with the above laws, regulations, guidelines etc. is checked by the commissioned organisations and institutions. This applies to the acceptance of the installation and the inspection of the ongoing operation. In Germany, the following institutions are essentially responsible for this:

- State supervisory authorities,
- professional associations, e.g. German Social Accident Insurance,
- official experts.

The development of safety specifications is made difficult by the fast-moving technology and the high variance of AGVs. This makes it all the more important to involve competent occupational safety organisations at an early stage of planning. The safety of AGVs is thus a task that concerns the vehicle manufacturer, the planning team and the operator in equal measure. Coordinated cooperation should be sought.

The existing guidelines must not only be taken into account in the realisation phase of the facilities, but already in the planning phase. Omissions can only be compensated for at a later date with additional effort. Therefore, the necessary vehicle and plant-related safety packages must already be detailed in the specifications. At the same time, the regulations, laws, directives, etc. to be taken into account must be listed.

In the case of AGV systems, the infrastructure for industrial trucks must comply in particular with DIN EN ISO 3691-4. This regulates, among other things, the requirements for the industrial truck (such as stability, warning systems, detection of persons in the travel path, emergency stop devices), testing of commissioning and operator information. Approvals for fire and explosion protection, immission control or building law may be required. This must be checked in each individual case.

At this point, we would like to point out that the AGVS project must achieve CE conformity that takes the entire environment of the AGVS into account. All interfaces, especially those to stationary conveyor systems, must be assessed by hazard analysis and integrated into CE conformity.

Acceptance

The limitation period for claims for defects begins with the formal acceptance according to the BGB.[15] Experience shows that acceptance according to the BGB only takes place after

- elimination of gross defects or defects that impair the function,
- submission of the CE documentation including the CE declaration of conformity for the entire installation,
- conduct performance and availability tests,
- submission of an agreed statement of additional/reduced costs; and
- all official approvals.

---

[15] BGB = German Civil Code.

The acceptance protocol created for this must be signed by all parties involved. The VDI Guideline 2710—Sheet 5 "Acceptance Specifications For Automated Guided Vehicle Systems (AGVS)", to which reference is made here, describes further special features of the acceptance of AGV Systems. It describes the scope of the acceptance as well as the procedure and the concrete implementation.

The appendices to the guideline are certainly interesting: Here there is both a schematic representation for determining the system availability and a form for an acceptance protocol.

5.2.4 Operational Planning

The aim of this phase is the careful planning of trouble-free AGVS operation. This includes involving the employees in the planning process in advance. All employees who will later have to deal with the system on a regular basis should be informed and consulted. Only then will they accept the driverless "colleagues"—a prerequisite for successful use of an AGV System.

In addition, selected employees must be trained and qualified to be able to remedy minor faults independently later on. These measures must be repeated regularly—especially for new employees.

These training measures also include mastering emergency strategies. These take effect when extraordinary incidents (such as fire alarms) occur, but also in the event of the failure of an important partial trade or sub-system to maintain production. These emergency strategies are very application-specific and must be defined in advance.

The following points still need to be planned, for which corresponding budgets must also be kept available:

Maintenance/Servicing
Maintenance includes all service measures that are necessary to maintain or restore the functionality of the system, in particular maintenance, inspection and repair. This includes the regular inspection and cleaning of the AGVs as well as the replacement of wear parts. The service intervals depend on the respective use of the AGVs by the client. They are determined individually by the supplier and should already be mentioned in the offer.

Maintenance can be carried out by the operator, the AGVS supplier or a third party. If the supplier or a third party takes over the maintenance, this is usually done via a service contract with an agreed response time. A service contract can be executed as a full service or partial service contract, depending on the customer's wishes.

Spare Parts Supply
In order to be able to quickly replace defective parts in the event of a malfunction, an adapted spare parts provision and supply is necessary. For this purpose, the risk of failure, susceptibility to failure, availability and supply capability are determined for each component. According to the assessment, the individual components are kept in different storage forms:

- Spare parts store at the operator with the most important components,
- consignment warehouse of the AGVS supplier at the customer's premises with the high-quality components (optional),
- supplier warehouse with all standard parts,
- subcontractor warehouse (OEM warehouse) with all components.

Safety Checks

The recurring safety checks in accordance with the accident prevention regulations (in Germany: UVV) must also be carried out by the AGV operator. If the operator is not in a position to do so, he must charge the supplier or an external service provider to do so.

5.2.5   Change Planning

This planning step deals with changes during the operation of the facility. These can be software and hardware updates required in the course of operation or replacements at the end of the service life, as well as optimisations or extensions to the system required in the course of operation, e.g. additional vehicles, extension or modification of the routes, integration of new transfer stations, etc.

The need for any change can come from the AGV system itself on the one hand, but also from the conditions of use on the other.

Need for Change, Starting from the AGV System

The rapid development in the field of data processing/control/microelectronics requires software updates, SW and/or HW upgrades and/or adjustments to the control system. After a certain time, spare parts and know-how are no longer available from the service staff. Then component discontinuations can also be expected.

Need for Change, Starting from the Conditions of Use

As professional as the planning and system designs were, there are always changes to the planning data. Production expansions or relocations can occur just as much as expansions of the AGVS' area of application. In such cases, the AGV systems' great hour has come: In contrast to most alternative transportation technologies, such system adaptations are technically easy to solve with the AGVS. In particular, they can usually be carried out without interrupting operations.

5.2.6   Decommissioning

This planning step concerns the end of the AGV System. The technical service life is usually between 10 and 20 years. It also depends on how the system is maintained and to what extent the control components are regularly updated. However, when the end is approaching, operational and legal requirements must be met.

The following reasons for decommissioning are common:

- System is outdated: Performance/availability/economy is no longer given.
- Maintenance and servicing can no longer be carried out rationally; retrofitting would be uneconomical.
- There have been gross changes in the area of application of the system, such as production task.

Before decommissioning, it could be checked whether the AGVS can be used in another area. Another alternative is to sell or pass on the vehicles to another AGVS user or back to the supplier. It may be possible to reuse parts of the system. Removed parts can be used as spare parts in another system.

In principle, in Germany the "Kreislaufwirtschafts- und Abfallgesetz" § 49 (KrW-/AbfG) must be taken into account when storing, transporting and disposing of systems and their components. The disposal of AGV components is usually unproblematic, as almost no disassembly work is required. Disposal can be carried out in-house, by the AGVS supplier or by any third party. In the case of disposal by the AGVS supplier, suitable concepts for recycling are most likely to be applied. The individual materials must be disposed of in accordance with the environmental compatibility guidelines.

The following applies to the disposal or recycling of batteries: Lead-acid batteries can be subjected to a regeneration process for reuse. Nickel-Cadmium and Lithium-based batteries are only partially recyclable. Both systems must be sent to special return systems for final disposal. Recycling or disposal is carried out in accordance with the current version of the Battery Ordinance.

5.3 Planning Support

More and more companies are discovering intralogistics as a rewarding field of activity for optimisations in the areas of quality, processes and costs. Personnel capacities and resources are being increased and experience with intralogistics projects is growing. On the other hand, there has always been a desire for information and support. We want to show here where one can find which kind of information and support.

The Role of Manufacturers During Planning

In the age of the Internet, it is easy to generate a long list of AGVS suppliers that the prospective customer has contacted within a very short time to request brochures and a quotation. Does the reputable AGVS supplier distinguish himself by promptly presenting a detailed and project-specific offer?

We have seen that AGVS planning is not trivial and must take into account a wealth of project-related boundary conditions and special features. If an AGVS supplier immediately starts asking about all these points and requesting further information, this would be technically correct, but would probably scare off one or two prospective customers. If he quickly comes up with an offer for an AGV control system, x vehicles and a fixed fee for the project-related services, he may meet the initial expectations of the enquiring prospective customer, but in the end he will not do justice to the matter.

In addition, the technology of automated guided vehicles is interesting to appealing for many engineers. This is certainly one reason for the fact that there are considerably more initial enquiries than realised projects. The AGVS supplier—especially the sales and project planning departments—must therefore economise with their resources. This is no easy task for the AGVS suppliers, especially since their product portfolio is more or less limited. The prospective customer will always ask himself whether the products offered are completely optimal for his project or whether they are only optimal for the supplier.

In any case, a reputable AGVS supplier has many years of planning experience. He can contribute this especially with regard to the following tasks:

- The basis of the work is the collection of framework data required for the assessment of the project. For this purpose, there is the VDI 2710-2, which is used by the manufacturers, adapted if necessary. Such data collections help the customer in particular to fix his planning start so that he can recognise or prove changes during the planning process.
- Choice of vehicle type: The loading equipment used by the customer as well as the layout and other criteria determine the type of load pick-up and thus the vehicle type. A typical question when transporting pallets is the choice between the automated pallet truck and a piggyback vehicle with lateral pallet pick-up by roller conveyor or telescopic forks. Both vehicle types have clear advantages and disadvantages.
- Calculation of the required number of vehicles: The number of AGVs needed is an essential prerequisite for the economic efficiency calculation. However, it also influences the traffic situation in the plant layout. For the calculation, the AGVS supplier uses the customer's transport and route matrix and incorporates his know-how about the movement behaviour of the vehicles into a static spreadsheet, which quickly gives him the utilisation of the plant depending on different numbers of vehicles.

VDI Guidelines

The Association of German Engineers (VDI) has had an expert committee on *Automated Guided Vehicle Systems (AGVS)* since the beginning of 1987 (named VDI FA 309). This committee has set itself the goal of strengthening the AGVS sector by bringing manufacturers and operators together and, as a neutral and recognised institution, bringing potential users closer to AGVS. It carries out fundamental guideline work with the aim of giving the industry more certainty for action and planning. In this way, the existing application possibilities are to be better exploited and new fields of application opened up (Fig. 5.3).

Well-founded and up-to-date rules and regulations should provide certainty. The following focal points serve this purpose:

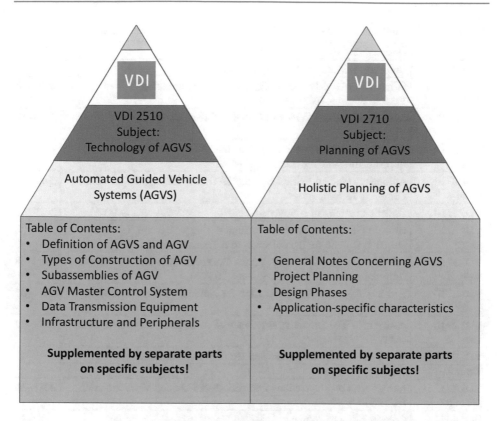

Fig. 5.3 The main topics of the VDI expert committee "Automated Guided Vehicle Systems"

1. Description of the current state of the art

    The existing rules and regulations are continuously adapted to the state of the art and further completed.

2. Creation of planning certainty

    The user gains certainty through the definition and description of holistic planning of systems. He receives assistance in the form of initial advice, basic information, guidelines, tools and aids.

3. Active market communication

    The AGV technology is made known to a broad public through targeted information such as events, publications, lectures, internet presence and promotional measures. This also includes the AGVS symposium,[16] which takes place every 2 years, since 2012 at the Fraunhofer Institute IML in Dortmund (Table 5.8).

---

[16] www.fts-fachtagung.org

Table 5.8 VDI Guidelines on the subject of AGV systems (technology and planning)

RL-No.	Designation	Date
VDI 2510	Automated Guided Vehicle Systems (AGVS)	2005–10
VDI 2510 Sheet 1	Infrastructure and Peripheral Equipment for Automated Guided Vehicle Systems (AGVS)	2009–12
VDI 2510 Sheet 2	Automated Guided Vehicles (AGVS)—Safety of AGVS	2013–12
VDI 2510 Sheet 3	Automated Guided Vehicle Systems (AGVS)—Interfaces to Infrastructure and Peripheral Equipment	2017–09
VDI 2510 Sheet 4	Automated Guided Vehicle Systems (AGVS)—Energy Supply and Charging Technology	2020–01
VDI 2710	Holistic Planning of Automated Guided Vehicle Systems (AGVS)	2010–04
VDI 2710 Sheet 1	Holistic Planning of Automated Guided Vehicle Systems (AGVS)—Decision Criteria for the Selection of a Transport System	2007–08
VDI 2710 Sheet 2	AGVS Checklist—Planning Aid for Operators and Manufacturers of Automated Guided Vehicle Systems (AGVS)	2008–08
VDI 2710 Sheet 3	Areas of Application of Simulation for Automated Guided Vehicles (AGVS)	2014–05
VDI 2710 Sheet 4	Analysis of The Economic Efficiency of Automated Guided Vehicles (AGVS)	2011–07
VDI 2710 Sheet 5	Acceptance Rules for Automated Guided Vehicles (AGVS)	2013–12
VDI 2710 Sheet 6	Introduction and Operation of an Automated Guided Vehicle System (AGVS)	2018–10
VDI 4451 Sheet 1	Compatibility of Automated Guided Vehicles (AGVs)—Manual Control Unit	1995–08
VDI 4451 Sheet 2	Compatibility of Automated Guided Vehicles (AGVs)—Energy Supply And Charging Technology	2000–10
VDI 4451 Sheet 3	Compatibility of Automated Guided Vehicles (AGVs)—Traction And Steering Drives	1998–03
VDI 4451 Sheet 4	Compatibility of Automated Guided Vehicles (AGVs)—Open Control Structure For Automated Guided Vehicles (AGVs)	1998–02
VDI 4451 Sheet 5	Compatibility of Automated Guided Vehicles (AGVs)—Interface Between Client And AGVS Master Control System	2005–10
VDI 4451 Sheet 6	Compatibility of Automated Guided Vehicles (AGVs)—Sensor Technology For Navigation And Control	2003–01
VDI 4451 Sheet 7	Compatibility of Automated Guided Vehicles (AGVs)—AGV Master Control System	2005–10

All VDI Guidelines can be obtained from Beuth-Verlag

Forum-FTS (www.forum-fts.com)

In 2006 all AGVS suppliers organised in the VDI committee FA309 founded the Forum-FTS[17] as a community of interests for the AGV systems industry. The group is currently made up of 20 members from five European countries. The Forum-FTS sees itself as a neutral contact point for (potential) users with a neutrally managed website.

The Forum-FTS presents itself at trade fairs and exhibitions and commits itself to a self-imposed code of honour in its dealings with each other and with customers. Ultimately, membership in Forum-FTS means a promise of quality. Our goal is successful AGVS projects!

The information and advice offered by Forum-FTS is not only helpful for inexperienced interested parties, but also for the experts. Because the market is constantly changing:

• the AGV manufacturers vary their profile,
• new technologies are coming onto the market and
• the world of technical regulations is alive.

A particularly popular offer is the initial AGVS consultation. For a moderate price, it offers initial statements on the technical feasibility and economic viability of the planned AGV project. It enables an efficient and neutral start to the project.

In the course of its existence, however, Forum-FTS has also developed into a planning and consulting company. At the latest from 2016, when Forum-FTS became a limited liability company (German: GmbH), it offers AGVS users every conceivable support for their AGVS project. For this purpose, there is not only the well-known initial AGVS consulting on the Internet, but many other concrete consulting packages that apply to all AGVS applications described above. Today, a team of five proven AGVS experts with more than 150 years of AGVS experience works here.

Today, Forum-FTS also takes over the activities that were covered by AWT-Kompetenz GmbH (www.awt-kompetenz.de) before 2016, i.e. intralogistics applications in hospitals and nursing homes. AWT is the (German) acronym for "automatic goods transport"—this abbreviation is used exclusively in hospitals and clinics. Modern AWT systems are often AGV systems.

---

[17] "FTS" (Fahrerloses Transportsystem) is the German acronym for AGV system.

Seminars and Training
Basic information on the topic of AGVS can of course be found in this book. However, there are a wide variety of additional formats for further training. For this, we would like to refer to the offers of the Forum-FTS where some suggestions are given. All kinds of seminars, lectures, training courses, workshops and specialist forums can be booked here. Individual topics, processes, venues and dates are of course possible.

Consultation and Planning (www.fts-kompetenz.de)
Last but not least, reference should be made to the management consultancy Dr. Ullrich, which specialises in the topic of AGVS and can be of interest not only to potential users (consulting, planning and implementation), but also to system or component suppliers (technology monitoring, market strategies). Insofar as user consulting is concerned, the services are now usually offered by Forum-FTS GmbH (Fig. 5.4).

5.4   10 Key Factors for Successful AGVS Projects

Finally, more than 30 years of personal AGVS project experience will be summarised. It will be shown why holistic planning is so important for successful projects and which points are essential for the AGVS project to be successful. All areas of planning with the most common planning mistakes will be summarised in 10 striking key factors. The direct address is aimed at the AGVS user!

5.4.1   Holistic Understanding of the Project and Design with Vision

Understand an AGVS as a solution, i.e. as a system, not as a limited number of automated vehicles. Because the AGVS points the way to the future for your own production and intralogistics. Often, at the beginning of a project, the only question asked is: "Can't this also be automated?" Especially with the status quo of in-house intralogistics with manual forklifts or a tugger train solution, this question can prove insufficient. Better is the visionary approach, namely the question "What should our production logistics look like in 5–10 years?" Only then you can come up with new structures, new layouts, new methods—perhaps with different types of vehicles (Fig. 5.5).

Fig. 5.4 The members of the Forum-FTS (as of 2019)

5.4.2   Technical Design vs. Technical Demand

Understand an AGVS as a great opportunity for a change into the future. It is important to understand the difference between "automatic" and "autonomous". The terms "autonomous robot", "autonomous vehicles" or "autonomous systems" instead of AGV (*automated* guided vehicle) are mostly used as buzzwords; autonomous functions are neither needed nor mastered in most intralogistics applications. We live in a time of hyperinflation of terms; you can even find "fully autonomous vehicles" on the internet (not in reality).

Fig. 5.5 The actual situation as a direct specification for automation?

The unfounded demand for autonomous vehicle functions can quickly lead to chaos, but above all to solutions that are not calculable in terms of performance. Essential in the technical design are the points of navigation and safety. Define exactly the required degree of flexibility of your intralogistics and choose the appropriate navigation or localisation method. Before the tender, learn which sensors are usually used for localisation, personal safety and, if necessary, machine or object protection. You should know how these sensors work in order to understand what to expect from each method.

5.4.3   Strong Specifications as the Technical Basis of the Project

The specifications are the most important document for a proper tender. Its careful preparation is particularly important because it is the most important contractual document of the project, along with the subsequent commissioning and the specifications.

When drawing up the specifications, it becomes clear to what extent the project has been thought through. Let neutral experts help you with this or at least have it checked before you distribute it to the manufacturers.

In any case, insist on a requirement specification that is prepared by the project manager of the supplier entrusted with the order. This specification is important because there are usually several ways to implement the specification and because the project manager of "your" project should show that he/she has understood the task and how he/she wants to solve it. Check the specifications carefully and release them when the project can start.

5.4.4   Project Managers with Expertise Hopefully on Both Sides

We are often asked whether a project manager with project experience and expertise is necessary, or whether a project manager from another field is sufficient because the expertise lies with the supplier. The answer is clear: an AGVS project is too important, the consequences of project delays or poor system performance are too serious for "one-sided competence" to be sufficient. Understand an AGVS project as a joint project of client and contractor; there should be expert project managers on both sides.

Experience shows that AGVS projects are often complex and fraught with problems during realisation, which can only be solved by a team of good project managers. For this, the project managers on both sides must be equipped with extensive competences.

5.4.5   Realistic Timetable with Milestones

There are net and gross periods; note the differences. Divide the project into phases and allocate net and gross periods to them. As the client, be realistic and not demanding, because reality will catch up with you anyway.

In this way, a realistic schedule is created, which is to be provided with milestones that include contractually foreseen interim acceptances. These interim acceptances serve as a continuous performance check (Fig. 5.6).

5.4.6   Integration of the AGVS into the Periphery vs. Adaptation of the Periphery to the AGV

Only demand as much from the AGVS as necessary and adapt the periphery to the future AGVS as much as possible. By periphery we mean the floor, the path widths, the other traffic, the load provision and load transfer, the container variety and the IT landscape. Do not demand too much from the AGVS, but use the opportunity to improve the overall situation in the factory in advance of the project with sensible adjustments to the periphery (Fig. 5.7).

Example floor conditions: If you know exactly that the floor conditions are currently not optimal, do not ask the AGVS or the AGVS supplier to accept the status quo as a given. Ensure that the ground is suitable for AGVs in advance and ensure that the ground is dry and clean during operation!

Fig. 5.6 A realistic schedule is the prerequisite for a serious project

Fig. 5.7 Order and cleanliness as a prerequisite and consequence of AGVS use

5.4.7   Early Integration of Occupational Safety, IT and Production

Ensure the timely involvement of the following departments or groups in automation considerations:

- *Occupational safety*, because they are confronted with factory regulations that can go beyond the legal (Machinery Directive, DIN EN ISO 3691-4, VDI Guidelines). Often the occupational safety is represented by employees with little experience and a lot of responsibility. The earlier these employees are involved, the less likely they are to intervene towards the end of commissioning.
- *IT* is placing increasingly high demands on the AGV master control system (currently almost exclusively proprietary solutions). IT must already include these requirements in the specifications and check during the specifications check whether the implementation of their ideas has been successful.
- The *production team* has to live with the system from the moment it is handed over. Here one can find human habituation processes that have jeopardised many a project. Involve the production staff in the planning process and give them the feeling that they can help shape the solution.
- At the client's premises, responsibility for the AGVS is often transferred from the planning team to the production team after acceptance. Here, the users and at least one super-user should be named who personalises the responsibility. These employees should be named at the very beginning and stay involved during the whole project.

5.4.8   Meeting Culture

The importance of an AGVS project and its complexity have already been mentioned. The duration of a project can easily last 12–24 months after the contract has been awarded. Therefore, a disciplined meeting culture is necessary to remember and prove appointments made months ago. This includes the following points:

- Start the project with an official project kick-off meeting. During this meeting, the transfer of responsibility from the supplier's sales department to the supplier's project management should take place. The project managers from both sides introduce themselves, get to know each other and agree on the way of communication as well as the schedule with milestones.
- Provide for scheduled regular meetings and have official minutes written by the contractor's project manager; these minutes should be filed as pdf documents to document the course of the project! E-mail or WhatsApp histories are unsuitable for subsequent tracking and for any evidence that may be required!
- Conference calls do a good job nowadays; however, only face-to-face meetings are suitable for reaching agreements on complex issues. If the project falters, if real

Fig. 5.8 Modern meetings: How to ensure the undivided attention of all participants?

compromises with concessions (no matter on which side) become necessary, insist on a face-to-face meeting!
- Neither smartphones nor tablets are allowed during the above-mentioned regular meetings! This is the only way to be sure of the undivided attention of all participants (Fig. 5.8).

5.4.9   Agreed Acceptance Procedures

Often it is difficult or not necessary at all to consider at the beginning of the project what the acceptance criteria will look like at the very end, i.e. during the acceptance of the delivered system, and how their fulfilment will be checked. In the specifications and in the functional design specifications, there is (hopefully) information on the performance and availability of the system. With reference to the relevant VDI Guidelines (VDI 2710 and 2710-5), the evaluation criteria and the testing methodology should already be described. In this way, one ultimately avoids different ideas regarding the duration of the tests and regarding the determination of performance and availability.

For the performance and availability tests, transport material in sufficient quantity and quality must be kept ready, also consider the space required for this! If these tests (have to) take place during ongoing normal operation in the production environment, consider possible effects on production processes. Also check early on whether a certain number of shift escorts should be agreed by the supplier.

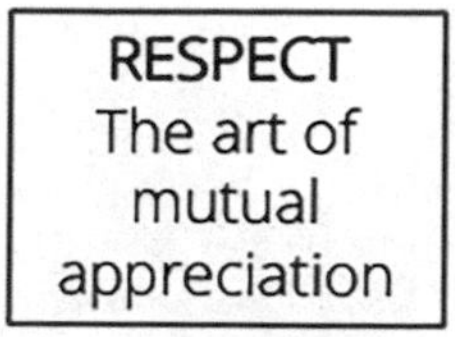

Fig. 5.9 Mutual respect as a
prerequisite in any technical
project

RESPECT
The art of
mutual
appreciation

5.4.10 Fair Dealings with Each Other

The AGVS supplier market has already been under strong pressure since 2017/2018. The requests have increased to such an extent that the classic AGVS market can no longer keep up. The providers can only process a part of the requests, the implementation times are getting longer. On the other hand, many new providers are appearing who are carrying out their first projects and have to prove themselves.

We are concerned to see a change in behaviour in our professional relationships. Fair dealings with each other should be a matter of course. The AGVS project is part of our professional tasks and thus a part of our lives. The project work should be based on honesty, openness, respect, reliability and professionalism. The "Ethical Principles of the Engineering Profession of the VDI" help with this.[18]

Only then any project can be successful and enjoyable (Fig. 5.9).

---

[18] Available for viewing and downloading on the VDI website.

Printed in the United States
by Baker & Taylor Publisher Services